War and the New Disorder
in the 21st Century

War and the New Disorder in the 21st Century

Jeremy Black

continuum
NEW YORK · LONDON

Continuum
The Tower Building
11 York Road
London SE1 7NX

15 East 26th Street
New York
NY 10010

www.continuumbooks.com

First published 2004

British Library Cataloguing-in-Publication Data
A catalogue record for this book is available from the British Library.

ISBN 0–8264–7124–2 HB

Typeset by RefineCatch Ltd, Bungay, Suffolk
Printed and bound by Cromwell Press Ltd, Trowbridge, Wilts

For Tim Rees

Contents

Abbreviations

ABM	anti-ballistic missile
ASEAN	Association of South-East Asian Nations
CIA	(US) Central Intelligence Agency
EU	European Union
FBI	(US) Federal Bureau of Investigation
IMF	International Monetary Fund
NATO	North Atlantic Treaty Organization
NMD	National Missile Defence Scheme
RAM	Revolution in Attitudes towards the Military
RMA	Revolution in Military Affairs
SALT 1	Strategic Arms Limitation Treaty (1972)
UN	United Nations

Preface

The end of the Cold War led to an attempt to define a new world order: a reconceptualization that was closely linked to particular political strategies in the 1990s, most vocally to the call for an order based on humanitarian norms and policed by peacekeeping forces operating under international authority. The validity of this interpretation of international relations was questionable, but irrespective of this it was placed under greater pressure as a result of the al-Qaeda attacks on New York and Washington on 11 September 2001. These led not only to a more militaristic attitude on the part of the American government and a related burst of war but also to a widespread lack of confidence that the new world order was more than an uneasily controlled disorder.

In the closing decade of the twentieth century there had been much talk of the obsolescence, not to say end, of war. Arguments varied, but a combination of the destructiveness and spread of atomic weaponry, a declining interest in both conquest and military service, and the supposed weakness, not to say collapse, of the state combined to lead to such claims. If war was outdated for these reasons it was also, in a separate analysis, presented as pointless because of the overwhelming military hegemony of one state, the USA, the leading economic and scientific power in the world. The notion of a Revolution in Military Affairs (RMA) was advanced in order to describe changes in the nature of military power and, in particular, both the technology of force and force projection, and the likely character of future warfare. In specific terms, there was an emphasis on information as a force multiplier, as well as a measure of military capability, and also as an objective in war: 'degrading' the information systems of an opponent was seen as the way to victory. As the USA monopolized the cutting edge of this technology and associated military ideas, it was seen as being a paradigm leap ahead of possible opponents.

In this book I take a different view, not least because I am unconvinced of the value of treating the leading military power as a paradigm force, rather than as the atypical state that its strength necessarily makes it. Throughout, I intend to try to take a global perspective rather than a Western one; and this will differ from most works on war, as these are very much American- and Euro-centric in character, inclined to take the Western state as a norm, and likely to focus on technological triumphalism. That was the prospectus outlined in my *War in the New Century*, a work written in 2000 and published in 2001, before the September 11 attacks. Others can determine how far I predicted the range and nature of warfare in the early years of the century, but my intention in this book is to take the welcome opportunity provided by the publishers of *War in the New Century* to write a longer study reprising its themes and re-examining its arguments in the light of subsequent conflict. As in 2000, but even more so as a result of the American interventions in Afghanistan and Iraq, there is the problem that Western warfare dominates general attention, leading to a failure to appreciate the diversity of conflict in the world, and its varied contexts, causes, courses and consequences.

In fact, despite claims to the contrary, these interventions have not altered the situation in which the notion of the RMA, and the conviction that 'smart' weaponry is the driving force in war and military capability, are misplaced, because they fail to incorporate the multiple contexts of war, not least differing understandings of victory, defeat, loss and suffering. In short, war and success in war are cultural constructs, and these constructs strongly interact with sociopolitical developments, so that, for example, democratization, changing gender relationships, and the resurgence of religion as a focus of identity and loyalty in much of the world, are more important than cruise missiles and other new weapons. Democratization, for example, has sapped willingness to accept conscription, while feminism, changing gender relationships and the role of women in armed forces are important in the decline of bellicosity in the West.

These will be the themes of Chapter 1, which will be taken forward into the rest of the book. Chapter 2 will focus on civil war, the rise of civil conflict and the decline of the state monopoly of force in given areas. This decline has international consequences, not least as collapsed or weak state forms, for example in Afghanistan, Congo, Liberia and Sudan, interact with international hostility.

Chapter 3 will look at the difficulty of sustaining systems of international agreement and mediation. Thus, in contrast to the claims of those who proclaimed a new world order of democratic capitalism after the fall of the Soviet Union, I will argue that such ideas reflected not a prospectus for the future but a degree of Western confidence, not to say hubris, and global influence in the 1990s that could not be sustained, not least because of the rejection of Western norms in China and large sections of the Islamic world, and the intractability of political problems and economic weaknesses across much of the rest of the world. The chapter will also address the issue of future conflict between the USA and China, and between the West and Islamic powers. In the latter case it is important to consider the continuing role of religion in international tension, and the Western, secular perspective that 'doesn't get' religion on its own terms, and therefore both fails to understand the motivations behind 'religious war' and is ill-equipped to negotiate exit strategies for these types of conflict. The issue of 'rogue states' will also be considered: it is important both in its own right and for what it says about the problems created for the West by the failure of the world to conform to Western norms.

Chapter 4 will pick up from there and will look at how conflict and weaponry are likely to develop in the future. Particular weight will be devoted to the interaction of and differences between land, sea and air capability and warfare. In the Conclusion, in place of the 'revolution in military affairs' approach, I will focus on the revolution in attitudes to the military and on the limitations of the technological approach.

As with all works, it is necessary to be wary of the degree of abstraction stemming from classification. A work on international relations presents states as building blocks, but that provides a false consistency and coherence in terms of apparent interests and views, because in practice national interests are not clear cut, either in the long or the short term, and debates over both interests and policy can be seen within states. This is particularly, but not only, the case with participatory democracies such as the USA and the UK. Mention of the USA serves as a reminder of the related point that the debate over national interests changes, both in response to domestic political developments and because shifts in the international system appear to dictate (at least to their advocates) changes in policy. The alterations in American policy in the 1990s and 2000s towards 'state-building' is an obvious case in point. The interventionism associated with the humanitarian liberalism of the Clinton government was replaced by a more cautious approach under George W. Bush and then, after the attacks on 11 September 2001, by a wideranging programmatic intervention focused on real or apparent threats but also linked to the notion that any failed state might become a cover for terrorism, the last an assumption with serious implications for diplomatic goals and military tasking.

As far as unwarranted abstraction is concerned, it is also the case that, although military forces are far more coherent than states, not least because they have formalized hierarchies, a culture of discipline, systems for command and a practice of control, nevertheless it is mistaken to suggest a consistency in capability and performance that is, in fact, often undermined by differences between units.

I would like to thank Robin Baird-Smith for proving a most helpful publisher, and Mark Bostridge for his valuable contribution as freelance editor. While writing this book, I have benefited from the opportunity to lecture on war in my War and the World course at Exeter University, which has provided a stimulating and friendly environment for the development of ideas, as have lectures given for the Strategy and Policy Division of the College of Distance Education of the US Naval War College at Pearl Harbor, Monterey and Washington, and for the Department of Continuing Education at the University of Oxford. I would also like to thank Bill Gibson for his comments on an earlier chapter and my son Timothy for helpful comments on the science of future soldiery. It is a great pleasure to dedicate this book to a good friend and much respected colleague. If this book leaves the reader, like the writer, feeling pessimistic, I hope it is a more informed pessimism. As far as my fears and predictions are concerned, I will be delighted to be proved wrong.

1 The Nature of War

| Either you are with us or you are with the terrorists.
> (President George W. Bush)

Writing in 2003, it is difficult to credit the claims made in the 1990s that war is obsolescent; or the suggestion that American power and new technology have made the outcome of conflict predictable. The notion that war is redundant, or at most an enforcement of the world order, appears today as a complacent echo of similar assertions made prior to World War I, when Western powers dominated the world and traded extensively with each other, and when systems of peaceful international arbitration were discussed and established. It is of course true today, as it was a century ago, that war can seem irrational, not least because a cost-benefit analysis of conflict rarely favours war, and this has become even more the case as the interconnectedness of the global economy has gathered pace. Furthermore, in a world of globalism, war truly can appear dated, for it depends upon a degree of self-sufficiency that is incompatible with the assumptions of modern political economy, with its stress on the free flow of trade and investment. However, such a policy of autarky, or at least of protectionism, may seem crucial to the maintenance of particular interests threatened by this political economy, while the assumptions of modern political economy do not generally enjoy deep public roots. Instead, populist attitudes frequently focus on antipathy, although that of course is far from identical with war.

There is a long tradition of wishing that war was anachronistic, a savage throwback to a former primitive state and/or an aberration within the course of human and societal development that could be overcome if only the impact of a range of supposed undesirables, such as capitalism or masculine assertiveness, could be educated away or otherwise eradicated. This hope was given eloquent expression on 3 September 1939, when Britain entered World War II in response to the unprovoked German attack on Poland and George Lansbury, the pacifist former leader of the Labour Party, told the House of Commons:

> The cause that I and a handful of friends represent is this morning, apparently, going down to ruin, but I think we ought to take heart of courage from the fact that after 2,000 years of war and strife, at last, even those who enter upon this colossal struggle have to admit that in the end force has not settled, and cannot and will not settle anything. I hope that out of this terrible calamity there will arise a real spirit, a spirit that will compel people to give up reliance on force, and that perhaps this time humanity will learn the lesson and refuse in the future to put its trust in poison gas, in the massacre of little children and universal slaughter.

War: an anachronism?

A belief that peace is natural and necessary, in the past, the present, the future – or all three – can be used to make war appear as unnecessary, but that does not necessarily make it anachronistic. Indeed, the very frequency of war, both during the last century and also the last decade, suggests that it is unhelpful to regard it as an aberration, and this is particularly the case in areas where conflict has been frequent, such as the Middle East and sub-Saharan Africa. If indeed 'war' is considered to include civil conflict and international quasi-wars then it can be seen as normative in many regions, indeed possibly the normal state of human society. Such an approach would see periods of peace as those of 'interwar'. Although literally true, this is a somewhat glib comment that provides no guidance to the extent to which peace is becoming more prevalent. One of the themes of this book is that war seems less normative and less normal in the Western world as the bellicosity of populations declines, that this is not the same as a fall in the frequency or impact of conflict across the world, and that the apparent disjuncture of these two developments can throw much light on a world that at once can appear more peaceful and yet be more violent.

There are particular reasons for an increase in tension and violence, aside from the general prevalence of violence as a means to pursue interests and assert identity. Globalism (and the characteristics of the modern world summarized by this phrase) is one of the major causes, as it accentuates the possibility of conflict for three reasons. First, the nature and pace of links within a globalized world create strains, especially in terms of economic pressures and living standards, that throw up a violence of 'counter-globalism'. This is readily apparent in the Islamic world, where 'counter-globalism', such as the destruction of instruments used to play secular music in Sudan, can be presented as a defence of values, and these living values are not rendered suspect to local opinion by being traditional. This potent rejection is directed not only at the international forces of globalism, especially multinationals, and foreign borrowing, their institutional structure, such as trade pacts and the International Monetary Fund (IMF), and their alleged global standard bearer, the USA, but also at local elites that are identified with this world. Hostility to globalism in many states means opposition to modernism and modernization. It can draw on powerful interests and deep fears, as well as express senses of identity, as in vandalistic attacks in France on branches of MacDonalds. Indeed, this hostility can also be seen in the USA, not only in terms of long-established resistance to aspects of cosmopolitanism and change but also in terms of political opposition to the constraints of international cooperation, the last most forcibly expressed by hostility to the United Nations.

Secondly, and conversely, the definition of states or groups that fail to conform to the international order and political economy of globalism, and to the political standards that are advocated, so-called 'rogue' states, leads to pressure on the stalwarts of this global order, especially the USA, to push them into line, either because of their domestic policies, or their international stance, or both. If this leads to military action, such a policy can be, and is, presented as policing and peacekeeping, but the reality is war. Indeed, the very willingness to consider, plan for and execute military schemes that are presented as peacekeeping suggests that war in that guise has become a ready response by liberal Western powers, such as the USA and Britain, that do not consider themselves authoritarian. The language of conflict is not simply indicative of its rationale but also central to its goal, because if it cannot be seen in these terms then support for the policy will be endangered: a problem that affected the British government over Iraq in 2003.

In the future, war in this guise may become a more common and widespread response to international disputes, but it may be as difficult to restrict the self-selection of peacekeepers to those who currently dominate 'peacekeeping' as it is to prevent the diffusion of weaponry from advanced states to others judged unreliable. This device of making military action appear more acceptable by defining it as peacekeeping is thus potentially destabilizing, and could cause war by emulation. This is in fact highly likely, as the global pretensions of peacekeeping are not matched by any marked effectiveness on the part of international institutions. Thus the ability of the United Nations to define peacekeeping and the peacekeepers, and so to legitimate, and de-legitimate, war is limited, and was seriously challenged by its sidelining by the USA and the UK in the Iraq crisis of 2003. Within some states, there is an analogous problem, as militarized peacekeeping by armies and police forces is rejected by some of the population, while the state authorities lack the legitimacy of consent, so that crisis at the international level can be matched at the domestic one.

Lastly, the very nature of a global economy is that it is both dynamic and prone to bring the far distant and very different into contact, which is in itself destabilizing. Thus, experience and expectations, 'real' interests and assumptions combine to make globalism a cause of conflict, however much this conflict undermines the logic and profitability of globalism, not least by introducing a powerful element of instability.

Aside from the issue of globalism, even if war is anachronistic for some, many or most states and interests, it is not so for all. Those who are willing to use force, for example Saddam Hussein when he invaded Kuwait in 1990, do not regard it as outdated, and such aggression can lead states that regard war as anachronistic to take part in conflict. In short, it is mistaken to argue from general developments, for example a decline in bellicosity in the West, to specific cases. In many senses, indeed, the course of both international relations and military history are the revenge of the specific on the general (and the generals).

There are, indeed, many groups that have found conflict welcome, and will continue to do so. It can seem the best way to pursue interests, not least if these are 'revisionist' (designed to reverse a past settlement). Less prudentially, it can have a cultural, emotional or ideological appeal that it is difficult to appreciate for those who do not share these assumptions or attitudes. It can also, and not least, be the salve for boredom or a sense of cultural ennui. In the past this encouraged a sense that military service or war were valuable cures for national ills: an attitude that continues to have more vitality than is sometimes appreciated. Even if deterrence, international pressure or pre-emptive or retributional action by powerful states can deter others from seeing war as the best way to pursue their interests, that does not mean that these forces can assuage non-prudential pressures for conflict.

Far from seeming anachronistic, war, defined as the use of organized force, in the first eight months of 2003, was the solution both of 'developed' states and of their 'less developed' counterparts. As far as actions by 'less developed', or powerful, states were concerned, it is possible to cite a number of conflicts, including those in Congo, Ivory Coast and Liberia, all civil wars. Further-more, as in these cases, such civil wars had an ability to draw in other powers, as, aside from attempts to exploit neighbourhood struggles, it is commonplace to accuse neighbours of sheltering rebels and to demand rights of pursuit; while refugees pose a major problem for neighbours. Whatever the response by neighbours, the struggle can spread.

It would be misleading to segregate 'developed' and 'less developed' states, because, aside from serious problems of definition with the term 'developed', it was also the case that 'developed' states tended not to fight each other but rather to employ their forces against the 'less developed'. This was true throughout the Cold War and the 1990s, but was pushed to the fore after the September 11 attacks introduced much of the newspaper-reading public to the concept of asymmetrical warfare. Although the sense that Americans were being introduced to the real world by these attacks was present in some of the commentary by those from other countries who had experienced terrorism, the destructive ambience and ambitions of al-Qaeda were different in kind to those of the other terrorist movements with which the West were familiar. In particular, the use of violence in order to achieve political goals, seen for example in Northern Ireland with the IRA, was different in type to the determination to punish the infidel which characterized al-Qaeda.

In practice, asymmetry is a matter not only of fighting but also of the wider conflict. Thus, the troubling strength and persistence of anti-Americanism in parts of the Islamic world requires a thoughtful and long-term political response that will contribute to the defence of America; looked at differently, destroying al-Qaeda will only profit the USA so much if another radical Islamic organization arises determined to repeat its policies and able to benefit from widespread alienation from the USA in the Islamic world.

At the same time, the focus on asymmetrical conflict increased in the 1990s and early 2000s because the extent of rivalry between major powers lessened as a consequence of the end of the Cold War and improved relations between NATO and formerly communist states. This was seen in American–Russian relations, although they were not without their difficulties, but, more dramatically, NATO was extended into formerly communist Eastern Europe, both with the expansion of NATO membership and the deployment of its forces into the region. The extension of membership is an ongoing process, so that, in November 2002 a NATO summit at Prague decided to invite Bulgaria, Estonia, Latvia, Lithuania, Romania, Slovakia and Slovenia to begin negotiations in order to join the alliance.

The current tendency of 'developed' states not to fight each other has underlined the focus on asymmetrical conflict, but is not necessarily some lasting rule or guide to the future, as, aside from direct conflict, they can fight or compete for influence through surrogates. This was very much the case in the Cold War, for example in the Middle East and Central America, and can also be seen in recent struggles, for example in Africa and the Caucasus, although the extent to which this is the case is a matter of dispute. It is not for example clear how far the civil and international wars in Central Africa, particularly in Rwanda and Congo, should be traced to French policy, as has been alleged, while it was also claimed by France that British or 'Anglo-American' interests lay behind Rwandan and Ugandan intervention in the Congo, but this has been denied. The extent of Russian involvement in conflicts involving Armenia, Azerbaijan and Georgia in the 1990s has also been a matter of controversy, but seems clearer.

The discussion of asymmetrical warfare helps in the understanding of war as a very diverse process: an understanding which has obvious implications for any analysis of conflict and more generally for the assessment of military capability. If wars are different, and military tasking therefore very varied, then it is unhelpful to think in terms of a single hierarchy of military capability, however that hierarchy is arranged, or to argue in terms of a narrow range of military characteristics.

War and technology

Currently, such a hierarchy is generally arranged in terms of military technology, particularly weaponry, although there are also other aspects of technology, especially those related to command and control functions, transport, logistics and medical care. These are all important – advances in food-preservation and water-purification, for example, being important to force projection. However, a focus on the material culture of war – whether weaponry and weapons systems or other aspects of technology – is only useful if it is contextualized by an understanding of the very varied character of war and of the extent to which superior technology does not necessarily bring victory, let alone success.

The latter is a lesson taught by military history, both distant and recent, and thus provides an opportunity to underline the importance of such history for an assessment of conflict and capability today and in the future. Such an assertion may appear anachronistic, and indeed military history on the whole plays a far smaller role in military education today than was the case a century ago: instead of considering past conflict, more time in military education has to be spent evaluating the consequences of technological advances for war at the tactical, operational and strategic level. The situation varies by country and service: air forces tend to be less interested in military history than armies and navies – a situation that owes something to the shorter history of air forces but more to the cultural assumptions of those who go into the service.

To fail to focus on modern technology would be foolish, as it is important to use weapons effectively; however, a wider understanding of the capabilities of military forces requires an assessment of conflict that brings out its multiple and unpredictable character and the range of other factors involved in success. This can be glimpsed by looking at military history in which there are many instances, the most prominent in recent decades being the unsuccessful American intervention in Indo-China during 1963–72. The latter is an extensively debated conflict, and there is much contention in particular about how far the American failure was due to the inherent difficulties of the task, and how far to its own domestic problems, ranging from an eventually critical media to excessive and/or unfortunate political interference in military policy. The two elements are sometimes linked by arguing that if the military had been permitted unrestricted bombing of North Vietnam or a ground invasion, the inherent difficulties of controlling South Vietnam could have been overcome. It is not the purpose of this book to re-examine the war, but current debates are part of the military world of the present day and will help to shape future assumptions; indeed, the reconsideration of the past is an important aspect of the current experience of conflict. In one respect, the reluctance by many who discuss the Vietnam War to accept the limitations of technology, especially air power, and the search for failure at home, including the constraints placed on strategy and operations, is the most important aspect of the debate, for it has obvious implications for how future conflict is considered. Furthermore, future war can be used to fight out different interpretations of the past, as with the debate over the effectiveness of air power.

On the global scale, the technological optimum, in terms of relative military capability, that was apparent in the late nineteenth century receded in the second half of the twentieth century. Conflict today and in the foreseeable future can be located in terms of this process of the erosion of military effectiveness and success stemming from technological advantage. As this is an important and in many respects counterintuitive point that underlies much of the analysis in this book, and which appears to clash with the lessons of recent conflicts, it requires some explanation. We may start with visual images, which have become more potent for us than for our predecessors because of colour and the incessant presence of film and television. The visual image of war on the global scale in the nineteenth century was of Western supremacy: close-packed lines or squares of Western infantry were depicted shooting down large numbers of non-Westerners, both infantry and cavalry, who hurled themselves towards the disciplined firepower of the Westerners. As with many images, this in part described a reality: there were battles like this – such as the British defeat of the Mahdists at Omdurman in the Sudan in 1898 – and there was an overall capability gap, but there were also engagements where

Westerners were less successful. Their weapons were useful, but issues of terrain, ecosystem, tactics, leadership, morale and unit cohesion were also important. In addition, modern weapons also spread to non-Westerners, being used for example to help Ethiopia defeat Italy at Adowa (Adua) in 1896. In any case, the advantage provided to Western forces by individual weapons could be lessened by the adoption of particular tactics by opponents.

War and politics

All of these factors ensured that the military situation in the nineteenth century was more complex than is sometimes appreciated. In addition, it is necessary to consider military operations within a wider political context in which Western imperialism appeared normative to Westernizers and could benefit from circumstances elsewhere, not least from the ability to win local cooperation, as in India, Nigeria and Central Asia. This context helped empower the Western military capability referred to, and should not be separated from the more narrow military analysis, because it is not secondary to it.

In the twentieth century, the politics of power changed, undercutting whatever military advantages the Westerners possessed, and also greatly altering their use in particular circumstances, and the manner in which their use was perceived. The politics, however, were the crucial change. Imperialism not only ceased to be normative in the West but also became unacceptable to those who were under imperial rule. These interrelated developments were both potent. They also did not arise at a particular moment, but instead were part of a shift that occurred over much of the century and continues today, providing the context within which military capability has to be judged. In short, political will is not a constant.

This shift is not stopping in 2004, nor due to stop in the foreseeable future. Indeed, in many respects, past, present and future are all linked here, because it is unlikely that we will move towards a situation in which imperialism is seen as desirable and acceptable. Part of the international criticism of American policy in 2002–3 related to a supposed imperialism that it was keen to disavow. This anti-imperialism is a fundamental building-block in the military situation, because military capability is in large part set by objectives, or 'tasking', and these are heavily influenced by the likely response to action. In other words, being 'fit for purpose' is a key aim for the military, but purpose is constructed in terms of particular political circumstances, both domestic and international; and this is true whether the military is subordinated to civilian political control, or whether it is autonomous, or even in control of the state. Thus, task-based doctrine has to take precedence over capability-based technology.

The Revolution in Attitudes towards the Military

Political circumstances have changed profoundly. Indeed there has been, and will continue to be, in the West a Revolution in Attitudes towards the Military (RAM) that has had a profound impact on the objectives and conduct of military operations and the nature of military institutions. This revolution has focused on a decline in the willingness to serve in the military, both in peacetime and in war, the causes of which vary by individual but can be summarized in terms of the movement towards societies that are more individualistic and towards a culture that is more hedonistic. Other 'isms' of the recent past and the present that are relevant include consumerism, capitalism and feminism, as well as democratization, each of which are more powerful in this and other contexts because of their interaction.

Feminism is particularly important, because it can be argued that there has been a reconceptualization of masculinity in the West in recent decades as part of a change in gender identities and relations. More specifically, the acceptability of constructions of masculinity in terms of the ability and willingness to be violent, and to be seen to be violent, and the heroism associated with self-sacrifice and death have fallen dramatically. This is important because much work on military effectiveness in combat conditions stresses the need to avoid shame in front of fellow-soldiers, and as definitions of bravery and shame change so conduct in the field is potentially affected. This situation is certainly a concern among opponents of the use of women as 'front-line troops', although, as Chapter 4 shows, the concept and practice of a front line have largely disappeared from modern Western military doctrine and warfare in favour of the notion of manoeuvre warfare, deep penetration advances and a zonal battlefield, so that any opposition to the use of women in the front line will in practice remove them from the combat sphere.

The acceptance of women for front-line units is a product of changes that result from feminism. Despite resistance in some quarters, this is likely to continue in the future, especially if fitness and other requirements can be appropriately varied, although it is important to note that in Israel, despite much myth to the contrary, women are not usually assigned to combat roles. The likely future rise in the number of women in front-line and command positions in many forces will not, however, lead to a presence comparable to men, because sexual stereotyping continues to be important and positive attitudes to the use of violence remain more characteristic of men. Nevertheless, many weapons are increasingly designed to take note of average female, as well as male, physical characteristics; while rising crime rates among women across much of the West suggest that gender stereotyping of attitudes to violence are increasingly inappropriate.

The role of women in the armed forces will also be an aspect of future military life that will remain heavily culturally conditioned. A major role for women will be far more true of Western societies than of Oriental counterparts, let alone Islamic states, and this will be especially true of command positions. The more prominent role of women will affect the image and self-image of Western forces, although there will be reluctance and resistance. Women may not play a major role in most non-Western regular forces, but movements with irregular forces are willing to accept such a role and benefit from the supposition that women are not generally fighters. Both Chechnyan and Palestinian terrorist organizations have used female suicide-bombers, the former, known as 'black widows', often the widows of Chechnyan fighters.

It is possible that talk about the reconceptualization of masculinity refers more to 'official' Western culture than to general social attitudes, and there are still important definitions of male identity and heroism in terms of violence, for example in film and video, especially in the USA. These frequently present violence as regenerative, reflecting a bellicosity that is powerful within Western society, although at the individuals' level social and legal pressures limit the use of violence in problem-solving.

Filmic images also capture an important shift in cultural values, because the heroes are often defiant individuals, loners who have to fight because of the supineness and folly, not to say treachery, of established channels of authority. In short, the discipline and comradeship of military units is not the model for modern images of exemplary violence, and these images conform even less to practices of command and to hierarchical structures. The stress on interdependability – and controlled inter-dependability at that – that characterizes both traditional and, even more, modern concepts of military organization and conflict is not the characteristic of the lone hero, whether on film or in a computer game. The notion of a revived civic militarism in the USA after the 2001 attacks discussed by some commentators has to confront this issue.

As an aspect of cultural shifts and social pressures, Western military forces are also being forced by political pressure to alter their attitude to the recruitment of pub-licly professed homosexuals, an issue that came to the fore in the USA in the early stages of the Clinton presidency. Because covert homosexuals were always part of the armed forces, this is less important a shift than the recruitment of women for combat units, but it is indicative of the extent to which Western forces are no longer able to set the parameters of military culture and increasingly conform instead to general social mores – a trend espe-cially apparent in the US military, despite the contrast between its overwhelming Republicanism and the far more divided views of the American public. Across much of the world, civilian control of the military today extends to the details of practice and ethos, which is further ensured by the intervention of litigation in internal military matters. This trend affects military structures and practices, including increasingly combat decisions, and it is likely to continue, as hitherto privileged jurisdictions are subordinated to the authority of judicial processes. The attempt by the Bush administration to avoid the purview of the American courts in the case of the prisoners held at Guantanamo Bay aroused a degree of scrutiny and criticism that indicated how far pretensions for military autonomy were suspected and opposed.

In part, this shift reflects the new definition of public interest in the West. In the past, such an interest would have protected military autonomy, but now to jurists such autonomy seems an abuse, or at least an anachronism, and their attitude takes precedence in government and society. It is also actively supported by media products (films and television) that harshly castigate supposed military abuses, such as alleged sexism or the bullying in military academies. To some within the military, the widespread acceptance by the public, politicians and military leaders of the desirability of criticism and judicial intervention amounts to a situation where it can seem as if preparedness is presented as less in the public interest than the pursuit of 'politically correct' strategies. At the very least, this opens up another contrast between Western militaries and those where there is no such attempt to reconceptualize (or, to its critics, attack) military culture, or at least military authority.

The end of conscription

The decline of conscription has been a measure of the RAM, although it is an imperfect one, both because conscription has had particular meanings in individual states and because conscription itself is not simply an index of bellicosity; indeed, as in Finland and Switzerland it can arise from defensive intentions, and is also important in societies that stress the idea of the citizen-soldier. This was true, for example, of revolutionary, nineteenth- and twentieth-century France, and more generally was seen as an important way to ensure the maintenance of accountability within the military and also democracy in society. This process was very important after World War II in countries that identified a professionalized military with right-wing politics: the position in several European states including Germany and Italy.

However, this civic militarism receded at the close of the twentieth century and looks set to continue to do so. It remains important in some states, especially Israel and Switzerland, while military service continues popular in Finland. However, civic militarism no longer frames the military culture of the major powers, nor of many other states that had previously emphasized conscription. Instead, conscription has become less acceptable in the West over the last four decades. Civic militarism has been eroded by social shifts and ideological and cultural pressures, and conscription has increasingly been seen as an aspect of authoritarianism and not, indeed, as civic militarism. Again, this is a good example of the extent to which changing conceptions that are not inherently military nevertheless have a direct and major impact on military life and possibilities.

Changes have multiple causes and consequences; thus over the last four decades a gulf also widened between conscription and military professionalism, and the latter was seen as more important in ensuring effectiveness than the numbers raised by conscription. The most important shift occurred in the strongest power, the USA, where conscription had been the legacy of World War II and was continued in peacetime during the Cold War. The Korean War (1950–53) underlined the USA's needs for numbers, leading to a revival of the draft, and it's worldwide commitments in the Cold War made the maintenance of conscription seem necessary. In addition, conscription very much accorded with the dominant social ethos of 1950s America. This ethos, however, was to be greatly challenged in the following decades when conscription fell victim to America's disillusionment with the Vietnam War, and with practices and symbols of authority. In 1968 Richard Nixon, then a presidential candidate, announced that he would get rid of the draft. He duly did this by 1973, a particularly important shift and example, as America was the world's leading military power and a key trainer of other militaries.

In Europe, Britain had already given up conscription, which did not have an historical purchase in public culture, and was only introduced to fight the two world wars. However, in most other countries, for example France and Russia, the decisive shift occurred in the 1990s, and was due for implementation by the end of the 2000s. In June 2000 the Italian parliament passed a bill phasing out conscription by 2005; it was designed to replace an army of 270,000 with a professional service of 190,000, including women. The Turkish army has begun to talk about abolishing conscription as part of a programme of modernization.

In states that preserve conscription, such as Germany, it was frequently discharged through social service. The special commission set up under Richard von Weizsäcker (a former president) to investigate the future of the German military recommended in 2000 that the number of conscripts be cut from 130,000 to 30,000, as part of a cut in the army from about 340,000 to about 240,000. Where conscription remains there is little effort to extend it to women, despite their new-found assertiveness throughout the West.

Such a shift away from conscription can be explained by arguing that these states did not face serious military challenges, and there is much truth in this analysis, but such an explanation is not the whole story, and the cultural changes already referred to were of great importance. So also were political factors. In Israel – a state that has repeatedly faced war and confrontation since it won independence in 1948 – conscription continued. However, in 2000 disillusionment with military service was given as a reason for Israel's abandonment of its presence in south Lebanon; while the willingness of conscripts to serve in the occupied territories on the West Bank and in the Gaza Strip subsequently became an issue.

In the case of Israel, and, indeed, other states where support for military service has ebbed, this shift can also be linked to a reconceptualization of such service, so that in place of the soldier as potential victim and an acceptance of the likelihood of casualties has come a search to banish casualty from the lexicon and an emphasis on an offensive–defensive stance focused on technology, especially air power. Thus, the Israelis became less keen on defensive positions, in which troops were vulnerable, preferring instead the idea of a reprisal military, essentially using the air to overcome the problems of the ground. This issue also affected the American forces in Iraq in 2003, exposing them to an unfamiliar context and putting a strain on morale. The apparent limitations of regular forces also encouraged the development of special units, for example the paramilitary capability of the Central Intelligence Agency (CIA), which was clearly seen in Afghanistan in 2001, as well as accentuating the development of unmanned drones.

Moves away from conscription did not reflect a uniform political or doctrinal situation. The situation was different, for example, for states such as those in Western Europe, where conscription was an aspect of territorial defence against threatening neighbours, and others, most obviously the USA but also the UK, where it had been linked to force projection. The break with the past element was also more pertinent for Eastern European states, in which conscription had been linked to communist control.

A move away from conscription was an aspect not only of individualism but also of the professionalization of war that became increasingly insistent as the large citizen militaries that had fought World War II were replaced across most of the world by far smaller regular forces. This again is a sphere in which the present seems to be no abrupt barrier between past and future but rather a continuation. This professionalization was questioned in the aftermath of the September 11 attacks, when calls in the USA for a revived civil militarism led to political interest in a reintroduction of conscription. However, this interest lacked political weight and governmental support and was certainly not welcomed by the military. Nevertheless, as an instance of the attempt to inculcate civil awareness and preparedness, a programme of smallpox vaccination was introduced in the USA.

Professionalization was assisted by the development of a branch of the military in which manpower was not crucial: air forces. A focus on machines, not manpower, had always been the case with navies, although they also required appreciable numbers of sailors. However, only a minority of states were naval powers, and, as a consequence, armies dominated models and practices of military organization and culture. Air forces offered a very different ratio of man and machine, man and killing power, man and cost, to armies, and their presence, role and aura assisted in the shift towards a more professionalized concept of war. So also did changes in land warfare. Changes that gathered pace in the last quarter of the twentieth century still retain their importance today, and will continue do so in the foreseeable future. The ratios referred to above shifted, as the effectiveness, cost and requirements on their operatives of particular machines increased.

The real cost of soldiers also rose, as the low-wage jobs of the nineteenth and early twentieth centuries were placed under pressure from rising wage rates and other costs, which were a function of growing affluence and expectations, especially but not only in the West. Costs per individual soldier had been held down under conscription, but that in fact increased the aggregate cost of the system, as large numbers of men, each of whom had little in particular to contribute, had to be equipped, transported, housed, clothed, fed and trained. All these demands absorbed much of the budget of armies that retained conscription, and reduced the amount available to spend on new weaponry. In addition, large numbers of fresh conscripts each year had to be trained, and from the most basic level, ensuring that conscript forces were very much unblooded barrack armies lacking mobility. Furthermore, as these soldiers mostly left military service quickly, the long-term benefit of such training was restricted, other than in creating a large potential reserve with limited and dated military skills. Conscription also seemed increasingly inefficient and obsolescent, because the requirements from individual soldiers for service in modern forces rose as training standards became higher and training regimes longer and more difficult in order to cope with complex machinery. As a result, the labour-market dimension of military recruitment was not that of the bottom of the market but rather that of competition in the skilled sector: a development that made retention a serious issue.

Thus, in opposition to conscription, cost and technological factors combined to encourage a professionalization that also reflected social trends. In short, militaries sought to transform themselves into volunteer forces, and these were seen as most likely to be truly effective, with effectiveness considered as much in terms of morale and unit cohesion as of equipment use. High morale and unit cohesion were regarded as important in fighting quality, as is indeed the case, and it is more difficult to develop and sustain such morale and cohesion in conscript units. There were also more direct political issues at stake: the high morale and sense of professionalism that in large part came from volunteer recruitment and long service also made it far less likely that military forces would reject or question political control and instructions that entailed the risk of casualties as demonstrated by the willingness of the British army to follow successive, and different, policies in Northern Ireland. Thus, in many Western states a citizen army in the age of democratization – a period in which there are increased demands for democratic practices and institutional responsiveness – appeared undesirable to governments, militaries and citizens alike.

Volunteer service meant that the military had to find ways to attract and retain sufficient numbers of the right kind of recruits. This led to a shift in the social politics and internal dynamics of military service, and was also related to both professionalism and improved military practice. In particular, a mutual trust of officers and men, based on a shared competence and reliability, helped to improve effectiveness. This looked towards future practice, as such a mutual trust became an objective of unit structures and training. Volunteer long-term service also led to greater concern about the families of military personnel, and about the impact of postings on family life, although provision for families increased the cost of the military system and diminished the ratio of expenditure on front-line preparedness.

The above comments are to some extent misleading because they adopt a paradigm – that of the Western army – and, having warned of the danger of creating a model and hierarchy based on Western technology, it then is inappropriate to do the same founded on a particular account of organizational development, specifically recruitment and internal dynamics. However, the processes described above are not restricted to the West. Indeed, there are important oriental parallels. In the 1990s the Chinese military embarked on a similar process of cutting numbers, and of professionalization and improved mechanization. The three were linked, although it is too simplistic to see the shift to different machines (i.e. a new technology) as driving the system. Military rather than social factors are, however, crucial to developments in China, because there is far less need to heed popular views than is the case for Western powers. A stress on regular forces, as opposed to mass armies and reserve militias, can also be seen in other leading Asian states, such as Japan and India. A million strong, the Indian military is very large, although it is small compared to the size of the population. The volunteer character of the Indian military is linked to a number of factors, not least the adoption from its origins under British rule of a Western model of professionalism and voluntary service. The absence of the resources necessary for conscription, or indeed for a larger military, is also important, as is military tasking. India wishes to be able to use its military to support paramilitaries in maintaining internal cohesion and seeks to appear and act as a regional force, but has no wish to conquer its neighbours, nor indeed to extend military experience to social groups judged unreliable.

The differentiation between the military and the remainder of the young adult male population that volunteer service produces is less marked in countries where state authority and the practice of impartial law and order are weaker, which is particularly true of parts of sub-Saharan Africa and of states such as Lebanon and Afghanistan. Here, the organizational account of differentiation between armed forces raised by voluntary recruitment and civilians who have chosen not to volunteer, seen in the West, India and Japan, conforms neither to long-established social patterns nor to recent political events. It is to these that we will now turn, not only because they are important in their own right, but also because they challenge any analysis of war and the military that focuses solely on state-to-state conflict.

2 The New Disorder: State Weakness

| **I can still punch.**
(Robert Mugabe, totalitarian President of Zimbabwe: interview
on South African television, June 2003)

As already suggested, in the intensely visual culture of the present, filmic images of war are the most potent, and with 'virtual reality' this is likely to become even more the case in the future. These images derive in great part from science fiction, where in the account of war seen in major films, such as *Independence Day* (1996) there is no comparison whatever between the two sides: one is human, the other alien. It is a fight to the finish and the finish is destruction: an echo of the absolute morality of the Cold War and a basic theme in science fiction. There are films and television programmes in which alien systems seek to coexist, for example the *Star Trek* series, but they are very much a minority.

In contrast, prior to the September 11 attacks, a depiction of other humans in terms of alien qualities and the inherent need for a fight to the finish (a metaphor for the Cold War) was no longer acceptable in what in the West was generally seen as civilized society – with the important exception of humanitarian interventionism against those practising genocide or 'ethnic cleansing'. In these cases, opponents were seen as denatured by their cruelty, and there was a conscious referral back to the Third Reich – the modern totem of horror – and a regime that provided an apparent parallel to the theme of good versus evil dominant in much of the world of science fiction.

The focus in science fiction on conflict between civilizations is analogous to the idea of sharply differentiated states, but even more there is an emphasis on technology and on war as occurring between manned (or aliened) weapons, with scant need to consider the civilian population. As a result of the focus on technology, science fiction appears similar to much of the discussion that surrounded the Strategic Defense Initiative (also known as Star Wars) plan announced by President Reagan in 1983, and, more generally, that around the current RMA discussed by many commentators, especially (but not only) in the USA.

In both the imaginative world, and in that of theorists and planners looking ahead to future conflict, there is an emphasis on the civilization/state and its armed forces as a unit actuated by technology, and as likely to be victorious as a consequence of its successful use. Most literature on the future of war would adopt this approach and be an exercise in futurology, which is not without value, for the military capability of the major states, and conflict between them (if it occurs), changes and will continue to do so. Thus, there is need for futurology, and we will turn to this approach in Chapter 4.

Civil conflict

However, such a perspective is far less suitable when it comes to looking at a range of conflict that in no way corresponds to the world of science fiction or 'high-tech' warfare, a range organized around two types of warfare that are far from coterminous but frequently overlap: civil war and 'low-tech' war. Civil war is the great forgotten in military history, a remark that might appear ridiculous given the amount of attention devoted, in particular, to the American Civil War (1861–65), and, to a far lesser extent, to the British civil wars of 1642–48; but the civil conflicts that receive most attention are those (like these) that oppose regular forces to each other, and can thus be readily accommodated to the central narrative in military history. In fact, this is not the case with most civil conflicts, with their prominent role for insurgents, guerrillas and terrorists fighting in less formal ways, and will not be so in the future. Their relative neglect is a serious problem in much discussion of war.

In addition, an account of future warfare that focuses on futuristic technology has little time for civil conflict, most of which is 'low-tech'. There has been some development of non-lethal weaponry that would be appropriate for dealing with hostile compatriots, but hitherto it has not played a large role in most discussion of such conflict, for although sticky substances and other incapacitants may be useful in controlling some riots it is difficult to see them as solving the problems posed by insurrection in, for example, Sri Lanka, or by terrorism in, for example, Peru.

It is probable that civil conflict will become more common, but prior to discussing this claim it is worth noting that civil conflict is difficult to define, even more so than international war. At one side, it fades off into policing: this is true of states with separate police and military forces, such as Britain, Spain and France, confronting violent separatism in Northern Ireland, the Basque Country and Corsica respectively. It is even more the case for states that lack a separate police force: there the maintenance of internal control is a prime responsibility of the military. Again, this is a division that should not be pushed too hard, for in many states the police, or part of the police, can be described as militarized; in others the challenge to ordinary policing is such that the army has to be used to maintain order.

This account sounds benign, in that the emphasis is on challenges to government and order, but it is also the case that in some states the prime challenge to order and good government, and indeed to society as a whole, can be presented as coming from the military itself, or from a state apparatus using force in order to maintain a control that is not answerable to the citizenry. Modern Myanmar (formerly Burma) and Iraq under Saddam Hussein are good examples of this. In such states, the brutal and frequent use of the regular army (as in Myanamar) or of militarized forces (such as the Republican Guard in Saddam's Iraq) are aspects of politics; indeed, civil conflict *is* politics, while politics entails conflict. The causes and nature of such conflict vary, but to look to the future is in part to look to the pressures that will exist within states and societies, to the methods of dealing with them and to the role and success of force in furthering strategies. Thus discussion of force and the military takes on some of the characteristics of 'total history'.

The nature and severity of the pressures that exist within states and societies have been regarded very differently over the last decade and a half, and there is no reason to anticipate any greater uniformity in the future. The fall of Soviet communism led to assertions of the so-called 'end of history' and more particularly of the end of divisions that had caused international and civil conflict. Thus a double hegemony was seen: on the global scale the 'new world order', presided over by the USA; and within states a combination of democratic politics and capitalist economics, both acting in a benign fashion.

Confidence in this projection ebbed during the 1990s and has not revived, as it became clear that there were powerful counter-currents, not least a resurgence of religious fundamentalism and of ethnic tension. In addition, an optimistic conviction of the ability to solve disputes short of conflict appeared less convincing than it had done in the early 1990s, as, more controversially, did the success of peacekeeping in solving, as opposed to containing, problems. Conflicts in the Balkans and Central Africa (but not only there) played a large role in affecting confidence, while the disruption created by global economic pressures became more apparent.

Demographic pressures

In considering the future, these factors are pertinent, while in addition there are structural changes in the world that suggest that more disputes will arise. The first relates to population increase, which provides a Malthusian vista of conflict derived from numbers exceeding resources, while also (as in Rwanda and Palestine) ensuring a large percentage of young men able to fuel conflict. While it is true that birth-rates in many countries are falling, and that aggregate global population growth is expected to fall after mid-century, it is nevertheless the case that the intervening growth is still seen as formidable. The 1999 UN Population Fund Study suggested a rise from six billion people in 1999 to 8.9 in 2050, figures that reflect not only the entry into fertility of current children, but also improvements in public health and medical care that lead to a rise in average life-expectancy, as well as the continuation in many countries of cultural restraints on restricting family size and the use of contraception.

This rise in population has tremendous resource implications, which will remain the case even if growth-rates slacken, because, aside from the rise in overall demand for employment and resources, there will also be a continuation in the rise in per capita demand. Indeed, belief in a likely fall in population growth-rates presupposes such a rise, as it asserts a virtuous linkage of economic growth and falling population. An alternative, predicated on rising levels of fatal diseases, is not anticipated, with the exception of AIDS in parts of sub-Saharan Africa, although the disruption caused by such diseases may well be a challenge for armed forces in the future. This is somewhat ironic, as troops have been a prime means of transmission of HIV in the region: the widespread conflicts of the 1990s and 2000s, particularly the intervention of a number of states in the civil war in Congo, ensured that the rapes and support for prostitution by which soldiers spread HIV were extensive in their geographical range.

Across the world, rising per capita demand is seen as a function, both cause and consequence, of economic growth and development, but these themselves are a cause of instability, because, despite important and continuing technological improvements in productive efficiency (e.g. the amount of water or fuel used in manufacturing processes) economic growth places major demands upon available resources. In addition, irrespective of economic growth, demand rises because of important social shifts, for example the move of much of the world's population into urban areas, which will continue, and which is linked to a decline in former patterns of deference and continuity, both within families and communities and more generally. In urban areas there is a willingness to reject parental aspirations and living standards, a decline in self-sufficiency and an increased exposure to consumerist pressures. However, such pressures, and the accompanying rejection of the past, are also present in rural areas, reflecting the role of migration as well as the massive extension of access to television in countries such as India and the place of advertising in modern societies. The destabilizing sense of a better world elsewhere that West German television brought to communist East Germany in the 1980s will be repeated throughout the world, and it is not surprising that Islamic fundamentalists seek to prevent or limit the spread of information about Western life, nor that the Western model is perceived as a threat by them, indeed a form of war, which leads to violent tension (for example in the northwest province of Pakistan in 2003) as efforts are made to suppress such signs and means of Westernization

Demands for goods and opportunities will be a cause of dispute and instability in families, communities and countries. Just as higher rates of unemployment tend to be linked to crime, although this can be cushioned by social welfare, and most of the unemployed are not criminous, so a sense of poverty, whether absolute or relative, encourages alienation and a feeling of dispossession. Violence is a response, although the reasons are complex, and definitions again pose problems. For example, squatting on rural land (as by the Movimiento Sin Tierra in Bolivia in 2003) or in urban property involves force, but should not always be regarded as a form of violence. Nevertheless, it is worth noting that in many areas affected by high rates of poverty, crime rates are high and/or rising. Thus, in Greater São Paulo in Brazil, the number of murders rose and remained above 6,000 in 1994 and above 8,000 in 1998. In 1999 a watershed was reached, with the rate of murders per 100,000 exceeding 50, as compared to 40 in 1994. In some areas such figures can be seen as a sign of a degree of social disorder that can be regarded as civil war, or at least warfare, as far as many individuals and communities are concerned. This is even more true in many states of the struggle against the drug trade, which destabilizes entire societies, leading to the corruption of politics and business and to high levels of crime and social breakdown. The use of the term 'war against drugs' in many states is a reminder of the reality and sense of struggle involved.

At the national level, rising demands for goods and opportunities will increase volatility in many states, and this will be particularly so in those that cannot ensure high growth-rates and the widespread distribution of the benefits of growth, or dampen or control expectations. There will also be an age dimension, in that ownership of property and financial assets will be concentrated among the older section of the population and spent on their welfare, to the dissatisfaction of the young. In some countries distinctions in ownership of wealth and access to opportunities stemming from large-scale immigration are also a major factor.

Demands for goods and opportunities will exacerbate problems of political management within (as well as between) states, encouraging the politics of grievance and redistribution. Conflict creates poverty, but poverty encourages conflict. Just as damaging is the threat of decline into poverty and of the relative poverty that will be felt by those who are comparatively well off but have not had their expectations realized. Such a condition can also be ensured if there are checks in the process of growth. Thus in Brazil, alongside the urban and rural poor whose hopes for a higher living standard (in some cases a living standard) have not been fulfilled will come those who are better educated and have gained jobs (for example in the state bureaucracy) but who feel angered by the limited benefits they have or can anticipate. This group may include sections of the military. Similarly, in Manila the attempted coup in 2003 in part rested on dissatisfaction among troops with their pay. At any rate, anger will be directed against those apparently benefiting from changing conditions and at their foreign links.

As a consequence, in many countries, economic growth may well not serve to assuage internal tensions, while there may well be no political or ideological cohesion within the state to encourage the elite to develop policies of sharing benefits or arranging welfare provision. As in the case of much of Latin America, the resulting tensions will interact with a hostility to the elite's modernizing ideology and policies, leading to an internal dissatisfaction that might easily escalate into civil conflict. This could take several forms (including violence against particular ethnic groups), but it will reflect the precariousness of government structures and the difficulty of developing systems of mutual benefit.

This schematic account is more true of some parts of the world than of others. By 1999, 95 per cent of the rise in the global population was occurring in 'developing countries', whose populations lacked adequate housing, sanitation and health services, and were also increasingly conscious of their relative deprivation. (In the case of Mexico and the USA this has led to individual migration rather than political discontent.) In addition, it was estimated in 1999 that nearly a billion people were illiterate, which again increases volatility. Furthermore, there was a political 'impoverishment', in that the means to press for significant change peacefully within the political system were often absent. This was an aspect of the problem of the failed state that was in fact far more common than the conventional use of the term would suggest. Indeed, from the perspective of this criterion, many states were failures, particularly those of the Third World.

Resource clashes

The resulting politics leads to grievances and clashes over resources, both of a conventional type, most obviously land and water, and of a more 'modern' type, such as quotas in educational opportunities, housing and government jobs, and the allocation of economic subsidies. Disputes over such issues make it easy for political groups to elicit popular support, and can make it very difficult to secure compromise, as they provide the lightning rod for regional, ethnic, religious and class tensions, for example in Pakistan. As instances of the resulting violence, the attempted coup in the Solomon Islands in 2000 arose from tension between two ethnic groups, the Malaitans and the Guadalcanal people, and the Malaita Eagle Force mounted the attempt in order to draw attention to its demands. Tension continued, leading, in 2003, to the formation of an Australian-led peacekeeping force. Also in 2000, tension over the benefits from the drilling of oil in the Niger delta in Nigeria led to growing violence as local people perceived few benefits flowing to them from their oil.

The potential danger of these tensions is readily apparent in some countries where grievances are freely voiced within the political process and also in the media. Such tensions are also present in states that make much more of an effort to suppress expression. The most powerful of these is China, where it is unclear how far the government will be able to contain the consequences of very varied growth-rates and economic conditions. Differences accelerated in the 1990s and early 2000s, not least with the expansion of special economic zones, and seem likely to continue to do so. In part, resulting tensions have been lessened by internal migration, but there are major problems. First, attempts by the central government to benefit from growth will lead to major rifts, not least because there is a powerful regional dimension, with expansion in the south and government in the north. Second, the resource demands of the rapidly expanding regions, particularly (but not only) for food and water will lead to resistance elsewhere as shortages develop and prices are pushed up. The range of possible conflict was indicated in Hunan province in 2000, as local television operators resisted attempts by China Telecom to enter the local television market with violence that so far has left 100 dead or wounded.

Furthermore, the Chinese military is increasingly fractured on regional grounds, with an identification between particular units and regional interests: a long-term factor in Chinese military history which reached its apogee with the warlord era of the 1920s. Communist attempts to use party control in order to overcome such fissiparous tendencies have had only limited success, not least because of the difficulty of rotating army units in order to break down their regional identification. Indonesia faces the same problem, and, indeed, it is a major issue in the military politics of large states, a branch of politics that generally receives insufficient attention.

It is unclear how far divisions within China will take on a military dimension. This is important for the future of warfare, not least due to issues of scale. The largest war in the nineteenth century occurred in China – the Taiping Rebellion – as did the largest (in terms of combatants) after 1945 – the Chinese Civil War. It may well be that in terms of people involved, the largest war in the twenty-first century will occur in China (currently the world's most populous state, although it is likely to be surpassed by India), and will be a civil war. This is a world away from the analysis of future warfare favoured by protagonists of a revolution in military affairs. Such a struggle requires consideration in any discussion of this future. The modern Chinese military is not well adapted for a sustained large-scale civil war, as modernization has led it to shed the numbers that might be required to hold down a dissident population, and, for example, escort food convoys. As a consequence, the war is likely to entail the use of police forces and maybe of militias, and will be presented to the outside world as resisting 'bandits'; an example of the Chinese method of managing news, which returns us to the problems of defining war. In India, such civil warfare – if it occurs – will probably have a religious and ethnic dynamic.

Such disputes can also be particularly brutal, which is in keeping with the general tendency of so-called primitive warfare to cause higher casualty rates than the majority of conflicts involving regular forces on both sides. For example, over the last half-century, smaller numbers have been killed by bombing than in close-quarter combat and (even more) wholesale slaughter. A recent example is provided by the very high death-rate in civil conflict in Rwanda, especially the Hutu-directed genocide in 1994, where many of the victims were beaten or hacked to death. Although regular forces can be vicious they are less likely than non-regular forces to engage in wholesale slaughter of ethnic and other groups judged to present a challenge, as can be seen with killings within Indonesia over the last decade, for example of Christians in East Timor and the Moluccas. The nuances of a moral hierarchy of discriminate versus indiscriminate killing is a matter for contention, but in Northern Ireland and Spain the indiscriminate killing of civilians was a characteristic of terrorists not of the army or the police.

Again, however, it would be unwise to adopt too clear-cut a distinction on this point. The conduct of the German army in World War II stands as a striking example of how a force that justifiably prided itself on its professionalism could play a role in genocide and large-scale brutality. There was not the shadow of a military reason for this conduct, and it cannot be extenuated by reference to the difficulties of holding down a hostile civilian population, for German brutality helped to encourage, if not cause, opposition. In many states it is difficult to distinguish between regulars and irregulars. In Congo, where maybe close to five million people have been killed in a multitude of overlapping civil wars from 1998, the militias that were responsible for much of the slaughter are difficult to classify, and an instance of the limited applicability of Western concepts. This is also true of the Tamil Tigers, who in their quest for a separatist, if not independent, homeland in Sri Lanka were responsible for large-scale violence.

Looking to the future, it is probable that there will be other episodes in which fear about ethnic difference will drive genocidal policies, or at least the savageries of 'ethnic cleansing'. It is also possible that the problems of operating in a hostile environment will lead to brutalities directed at civilians, not least because of the difficulties of identifying guerrillas from the rest of the population; the last, at the small-scale level, apparent with the problems facing a conscientious and trained military in the shape of the American army in Iraq in 2003. The situation is more savage for militaries that lack this ethos and training, especially if they are alienated from the people among whom they are operating for reasons of ethnic hostility, as for instance when Saddam's forces operated in Kurdistan.

The prospect of chaos and conflict over resources is not restricted to poor countries, nor to those with non-democratic political and governmental systems. For example, in democratic India there is a major contrast between the forces of the central government, which include rockets capable of carrying nuclear warheads, and the situation in a state like Bihar, where the private militias of landlords, especially the Ranbir Sena, compete with Maoist Naxalite guerrillas, and the latter compete with each other. Such competition can take on the intensity of a low-level war, and intercommunity violence is not easily suppressed by small high-tech regular forces. Were more states to have nuclear weapons, as is likely to be the case, the contrast would become more common.

Resource issues have in part been lessened over the last century as a result of economic growth and technological improvement. For example the 'green revolution' in agricultural yields has been particularly important in enabling much of the Third World to feed itself and the First World to do so without having to use its purchasing power to take too much food from the latter. However, it is by no means clear that such growth can be maintained, certainly at a high enough level to distribute food so as to meet individual and aggregate demand. This is particularly serious in areas of rapid population growth such as India, and indeed elsewhere, given the ability of hungry regions to export their hunger: by importing food if wealthy, or exporting migrants if poor. Thus, alongside failed states are famine societies. As far as food is concerned, the assumption that science will always provide is less sure today, as confidence in both science and rationalist solutions ebbs. This is the backdrop to the debate over the use of genetically modified crops, although it is possible that such crops will produce another green revolution. If crops can be modified to require less water and fewer pesticides then the economic and environmental benefits will be considerable.

The environmental damage of industrial society and modern agricultural practice is increasingly apparent, and although governments eager for economic growth, such as those of China, India, Indonesia and Malaysia, tend to ignore such damage, it may indeed feed into the general picture of disorder more directly by undermining growth and challenging stability. For example, the denuding and contamination of ground and underground water supplies is a serious problem, affecting agriculture in India and elsewhere. As with the loss of soil through deforestation and inappropriate agricultural regimes it will cause enforced migrations which will in turn interact with volatile domestic divisions. For example, concern about water supplies will exacerbate relations between groups practising different forms of agriculture. There will be claims that minority groups are responsible for the contamination or depletion of environmental resources, as with the high rates of water use by Jewish settlements on the West Bank of the Jordan. This will be viewed as a form of internal conflict, although Israel has in fact pioneered the efficient system of drip irrigation. Across much of the world, concern about water supplies is frequently cited as a reason for opposition to immigrants or incomers, which is another form of civil tension.

Around the world, resource problems will not prevent high levels of military expenditure, but may play a role in helping direct its expenditure, both by leading to conflict and by encouraging attacks on nodes in resource systems, for example pumping stations, power stations and dams. Terrorist groups attacked pipelines in Iraq and Colombia in 2003. In the same year the vulnerability of resource networks to attack was exposed when the triggering of a shortfall in electricity generation led to a dramatic failure of power in the USA. Although this was not due to terrorism, the crisis indicated a vulnerability that could be exploited.

Resource issues will be a major problem not only for poor states but also for the leading economic powers, reflecting as they do a rise in real prices and limited capability for efficiency gains through product substitution or new production techniques. While this will not be true for all products, as the price of oil over the last three decades has shown, or indeed for all of the time, it will be sufficiently so to be an important factor. A recent precursor was the pressure in 2000 created by rising oil prices, which caused domestic political problems in a number of

states, both 'Third World', such as Zimbabwe, and Western, including the USA, Britain and, most seriously,
France. In 2003 the end to government subsidies of oil in
Nigeria (a major oil-producing state) led to a general strike
and large-scale violence, with police firing on rioters,
followed by a government climbdown.

There will be future instances of resource shortages, and
they will create numerous tensions, both domestic and
international. Energy costs will be a particular focus of
tension, not least because of the failure of nuclear power
to fulfil expectations and the absence of a sequential
development in the nuclear field. As a result, the focus
will continue to be on the exploitation of nonrenewable
resources. Rising costs will encourage the exploitation of
hitherto unprofitable resources, but this process will not
provide low-cost energy, and once tapped the resources
will also be rapidly depleted. Oil production and reserves
were an important reason for rising American interest in
sub-Saharan Africa in the early 2000s, and for Chinese and
Japanese competition in Iran in 2003. They were also seen
as a major factor in the American attack on Iraq in 2003,
although this was denied by the American government.

As inexpensive and immediately accessible energy has
been internalized and made normative in the assumptions
of Western life, especially in the USA, its removal will
cause great stress and lead to a demand for action and a
search for culprits. This will be particularly true from
economically marginal groups whose living standards are
already under serious pressure, such as the rural poor, but
such 'marginality' may extend to all those whose socioeconomic position is precarious, including industrial
workers and others threatened by global competition.
Indeed, much of the population, including the military

and their families, may be affected. A limited willingness to accept the consequences of global shifts in energy prices and the power of the state to set tax levels was readily apparent in France in 2000 and Nigeria in 2003, and the political process will channel such anxiety and anger, leading to confrontation and conflict, expressed in terms of extremist political movements, contentious internal regulation and international disputes. In short, resources may be the cause and/or occasion for conflict whether the state, region or social group that is the aggressor is experiencing economic growth or the reverse. At a more minor level, the large-scale pirating of production, for example of pharmaceuticals, is in part evidence of resource tension.

There have been many accounts of why economic growth and social development will cause peace. In 1828 William Mackinnon, later a long-serving MP, claimed, in his *On the Rise, Progress and Present State of Public Opinion in Great Britain and Other Parts of the World*, that

> **the prevalence of public opinion may be the cause of hostilities between nations not being so common as in days of ignorance . . . when the individuals of those classes that most influence public opinion are aware that the pressure of taxation will be felt, more or less, in consequence, the community will not permit themselves, as in former days of ignorance or barbarism, to worry and attack their neighbours for mere pastime, or to gratify their caprice or warlike inclination . . . To argue, that because war has desolated Europe almost without intermission, longer than the memory of man or history can record, it will be as frequent in future, would be judging erroneously, and not**

> making sufficient allowance for the present state of
> civilization and power of public opinion . . . As other nations
> become civilized . . . communities will be benefited, in
> general, by an interchange of commodities . . . As
> civilization extends itself, the art of war is brought to
> greater perfections, and the burdens attendant on such
> warfare press more heavily on the community . . . In an
> improved commercial and agricultural state, wars are
> seldom undertaken but for the sake of preserving
> independence, or of obtaining some great commercial or
> political advantage; as they necessarily tend to impoverish
> the community which governs itself by public opinion, and
> acts according to its interests: hostilities, therefore, are not
> likely to be undertaken hastily, to be waged with acrimony,
> or extended unnecessarily.

This comforting account left, however, no room for ideology and politicization. In the event, the rise of commerce,
industry and the middle classes did not prevent bitter
and costly wars between 'civilized' states, nor indeed the
brutal conquest by Westerners of much of the world that
Mackinnon thought barbarous. Ironically, one of his own
sons was to die in the Crimean War.

Looking to the future, it might still be argued that prosperity and the pursuit of profit ensure that war will be
unlikely for advanced states, because the benefits of ease
will be widely distributed. Such claims have indeed been
made. However, it is just as likely that the segregation of a
professionalized military and the cult of new technology
will encourage the notion of a costless war. In practice,
as the American attack on Iraq in 2003 showed, however speedy it might be, such conflict is very costly,
and likely to remain so. The very essence of a modern

professionalized military is that it requires specialized equipment and support systems that cannot be readily improvised or requisitioned from civilian uses, although in the case of shipping there have been attempts to encourage convertibility and thus to lessen the need for specialist production runs. Nevertheless, there is greater specialization than was the case in the pre-machine age, and as production runs have become shorter so the real cost of military hardware has risen. The Iraqi conflict also showed that costs remained high after the manoeuvrist conflict was over, and indeed after President Bush had declared the end of hostilities. In July 2003 the cost of keeping the 145,000 American troops in Iraq was revealed to be $3.9 billion a month, while a steady number of American soldiers were killed in terrorist attacks.

Any sustained high-tech war between major powers would be very expensive, but it is unclear whether the factors mentioned by Mackinnon would in fact prevent conflict. The principal domestic sufferers in any such war would be those affected by cutbacks in social welfare and by wartime inflation (e.g. the elderly), neither of which group consists of the more 'empowered' members of society. Irrespective of this, the mechanics do not exist in crises, especially in the rapid run-ups to conflict (even if delayed, as with the two Gulf Wars), to allow the careful evaluation of popular views and in particular the extent of a disinclination to fight, if indeed such exists. Furthermore, foreign policy and defence are areas of policy that remain particularly under the scope of government control, and in which secrecy is defended. This makes it far harder to conduct any informed policy debate and to point out the drawbacks and unpredictabilities of conflict. This was readily apparent during the Kosovo crisis of 1999, and even more so in the Iraqi crisis of 2003. It is unlikely that this situation will change. Government attempts at news-management (i.e. 'spin') accentuate the issue of the disclosure of secret information, and further diminish the chance of informed debate.

States under strain

Resource competition is not the sole structural issue causing conflict between and within states. Another is that posed by the breakdown in the integrative capacity of states, which can be seen not only in the countries generally presented as facing problems of widespread poverty but also in some of their affluent counterparts. In general, assimilative ideologies and practices have become weaker (or, in many cases, remained weak), and governments find it increasingly hard to persuade minorities to renounce violent opposition, as with the FARC guerrilla movement in Colombia. Many governments also find it difficult to restrain their own propensity to use their authority in order to advance policies without sensitivity to the interests of sections of society. This is a problem not only in autocratic societies but also in democratic states that use a majority mandate to ignore minority views, or which indeed are also themselves faced by a violent minority unwilling to heed the democratic process.

In theory, modern states will be far better able to control and suppress discontent, as they will have the capacity to create a surveillance society in which the government will possess considerable information about every individual, including their location (the last offered now not by the census or identity papers but by satellite surveillance of cars). Furthermore, the nature of the modern salaried workforce and (through social security) non-workforce is such that most people will not be able to break from this surveillance society. It is possible that information capabilities will grow if in the future chips are implanted in individuals. There will be social benefit excuses for this, but the consequence will be to enhance the surveillance resources of government.

Aside from information, Western governments will also continue to have important resources. Their internal control forces, whether military or civil, will have communications and command and control facilities that will exceed those enjoyed even recently, and will continue to be more effective than those of domestic opponents. Furthermore, training for dealing with internal control and the provision of specialized units will both improve. Thus, Alvaro Uribe, the Colombian president, was able to complement the extra police and the part-time peasant soldiers he sent against the FARC in 2002 with a helicopter-borne rapid-reaction capability, while the USA provided both training and information from spy-planes.

Yet, as with the emphasis on technology in international military capability and conflict, such a stress on its role in internal control can be misleading, and can also risk extrapolating to the entire world circumstances that will only pertain in part of it. Furthermore, the factors of individualism noted in the discussion of RAM can be extended to internal control, for a breakdown of respect for government will lessen the ability of state agencies to elicit, shape and contain movements within society. In another important social shift, a process of social atomization may well leave widespread enthusiasms as the major way in which activists are motivated, and these enthusiasts will in some cases be unwilling to accept the disciplines of citizenship, not only in non-democratic states but also in their democratic counterparts. Subordination to majority opinions, and mutual tolerance against the background of the rule of law, will entail a degree of restraint that is increasingly unwelcome.

It is unclear that states, whatever their capacity for surveillance, will be able to contain the consequent tension, nor suppress the resulting violence. This is partly due to the difficulty of the task and partly a product of the constraints affecting their response. The balance of acceptable restraint will continue to move against the forces of the state and will be enforced by legal systems that frequently have scant understanding of the problems of acting against terrorists. Most Western states would have been unwilling to respond to a terrorist takeover of a theatre as the Russians did in Moscow in 2002, when their use of gas followed by an assault led to the death not only of most of the Chechen terrorists but also of 129 hostages. The international context also provides parameters. Thus, the choice of measures in the fight against FARC in Colombia is in part affected by the expectations of the American government supplying aid.

The difficulties of dealing with terrorist acts will be compounded in some countries by the political willingness to buy off terrorists with amnesties – a process that offers terrorism respectability, encourages further violence and undermines peaceful opposition movements. Such a process also defies the sense of justice that is important in encouraging popular consent and in the effective operation of democratic societies, although the situation is different in the absence of democratic processes or in separatist struggles where terrorism is sometimes supported by at least some sections of the community as an expression of their views. The latter is a problem for the Russians in Chechnya, as the absence of a political strategy to win 'hearts and minds', or indeed even serious encouragement to the military not to brutalize the population, lessens the value of the use of force against terrorists. The lack of any Russian political strategy capable of serving as the basis for negotiations with at least a significant portion of the opposition has both exacerbated and been exacerbated by the military situation, while the profit made by elements on both sides of the conflict (for example officers running protection-rackets) adds further problems.

As a consequence of the widespread difficulty of repressing disorder, peacekeeping as a military task will interact with what has been seen as the breakdown, or at least reconceptualization, of the state: a concept frequently expressed by referring to the 'post-Westphalian' state. (This is the transition of state authority away from the sovereign state and its monopolization of sovereignty that stemmed in Europe from the Peace of Westphalia of 1648, which concluded the Thirty Years War, towards a more diffuse situation in which national states wield less authority and sovereignty is multiple.) The weakness of government will be linked in some states to the presence of insurrectionary movements. Thus, in Colombia in 2002 the government's writ scarcely ran in half of the country. This issue was particularly acute in the case of countries with separatist movements, which affect a number of states from the very large, such as India and Russia, to the far smaller, such as Moldova, which has a Russian-backed secessionist section called Transdniestria. Whether or not the target is separatism, counterinsurgency warfare in response will pose a major challenge for militaries, both high- and low-tech. The principal problem will be the political one that plays a role in all counterinsurgency struggles: how to translate military presence into an agreed solution that permits a demilitarization of disputes; and this is more important than any differentiation of counterinsurgency strategies in terms of weaponry.

A geographical perspective can be added to this by suggesting that the next century might see a recurrence of the position in Europe in 1550–1650. Then, during the so-called Wars of Religion, conflict was most serious and protracted in areas where there were appreciable numbers of both Catholics and Protestants, rather than, for example in Iberia or Scandinavia, where one group predominated. Thus in the future the breakdown of integrative patterns will make multiethnic and multireligious states especially volatile, if the ethnic and religious groups maintain both a strong tradition of group cohesion and hostility to counterparts. This is at once pessimistic and in need of qualification. As most states are far from homogenous, the implication would be that high levels of civil violence are, and will be, present, but the former is not the case because practices of prejudice and discrimination, as well as the maintenance of cohesion through endogamy, confessional education and inheritance and employment practices are not the same as organized violence, nor do they necessarily cause the breakdown of the state. However, as Nigeria, Northern Ireland, Rwanda, Sri Lanka, Yugoslavia and other states showed in the late twentieth century, these practices retained this potential, and this is likely to continue to set the pace of civil conflict over the next century.

Religious conflict

The growth of religious fundamentalism has been a widespread 'structural' change, creating tension and a sense of cultures or nations under threat. Religious fault-lines are particularly powerful at the edges of the Islamic world, especially in South Asia, the Balkans and sub-Saharan Africa. In 2000, for example – and these instances can be multiplied – disputes between Christians and Muslims led to bloody riots in Kaduna in northern Nigeria in which about 2,000 people died, as well as to religious conflict in Indonesia's Molucca Islands in which some soldiers took a role. An alleged newspaper insult to the Prophet Muhammad linked to the Miss World competition led to new riots in Kaduna in 2002 in which over 220 people died. The sense of northern Nigeria as a religious frontier was particularly strong among both communities. Also in Indonesia, in 2000 there was separatist and interreligious violence in the Aceh region of Sumatra, and this revived in 2003, when the breakdown of a ceasefire led to an offensive by the Indonesian army. In the Philippines there is no sign that the long-lasting conflict between the government and the Moro Islamic Liberation Front on Mindanao will end, and, indeed, in 2000, the Front's leader, Hashim Salamat, called for a jihad.

Religious differences challenge not only the peace of both Indonesia and India but also their identity, as the secular ideology of the post-independence period is supplanted by religious-based notions of national identity, respectively Muslim and Hindu, marginalizing other religious groups. This is an important clue to future instability that will have international as well as domestic consequences, for such shifts make it difficult to accept compromise solutions to disputes. An example of this is the role of Hindu nationalism in the Kashmir dispute: a settlement that acknowledged the right of the Islamic majority in Kashmir to determine their own future, thus lessening tension with Pakistan, is unacceptable to these nationalists. The ability of religious groups to focus and exacerbate other tensions and to challenge the state is not restricted to Islam or Hinduism. In India it can also be seen with Sikh separatism in the Punjab, and in Africa with Uganda's Holy Spirit Movement.

As the resilience of religion, despite secularist analyses, assumptions and policies, has been one of the major themes of the last decade, it is likely that religion will remain important to the agenda and context of domestic politics, and indeed international relations, over the next century. Religious identity and antagonism helps overcome restraints against violence, especially (as seen in the former Yugoslavia) against the killing of neighbours, and also do not lend themselves to compromise. Furthermore, thanks to both proselytism and different birth-rates among religious groups, for example far higher rates in Gaza or among Kosovans than among Israelis or Serbs, confessional relations are necessarily dynamic or to use a different but equally appropriate phrase, unstable.

As a consequence, it is likely that religious antagonisms will play a major role in civil conflict over the next century. Prior to the September 11 attacks, this was underrated by most Western commentators, who are overly apt to identify religious antagonism with the past and particularly not with the Western world. The attitudes of most Western religious leaders helps support this approach as they stress ecumenical approaches, while, in addition, both theology and religious teaching in the West searches for common themes among religions. Yet these approaches are not only a misleading description of the potential of religious hatred throughout the world but may also underrate the likely role of such hatred in the West. In particular, church hierarchies and those following established (mainstream) Christian practices may find in the future that religious fundamentalism has a greater impact within Christianity. This fundamentalism might be directed at less ardent Christians as well as non-believers, and, as is all too frequently shown, movements of religious renewal do not have to be benign. The failure of many states to meet expectations or needs for identity or security encourage such movements, but the combination ensures that the most potent senses of identity are not coterminous with the citizenry.

Internal challenges

The failure of many governments to monopolize weaponry, to delegitimate internal violence and/or to control their own states does not have to lead to civil warfare, but it is very serious when linked to the absence or demise of consensual politics; although, looked at differently, this absence or demise encourages this failure. The net result is civil war of some type, although this can be variously defined. It may be no more than the armed defiance of the state seen with powerful gangs practising 'narcoterrorism', which is a major problem in Mexico, some South American countries (especially Colombia), much of the West Indies (for example Jamaica) and several countries in South Asia. 'Narcoterrorism' is likely to grow as the profitability of the drugs trade will remain high. Such practices offer effective tax evasion, and it is possible that narco-terrorism will create a greater challenge elsewhere, both in the West and in the Third World.

There may instead be some significant ethnic, religious, regional or political dimension to civil warfare, but the two may overlap, as with the FARC and ELN guerrilla movements in Colombia, both of which use force to hold sway over drug-growing areas and profit from the trade, as the Viet Cong also did during the Vietnam War. With the 'war on drugs' policy, the American government seeks to overcome the conflation of guerrilla movements with the drugs trade, but, paradoxically, it is American demand for drugs that finances this nexus. The American state uses methods against drug-producers and dealers in Latin America (and adopts a militarized stance to this end) that it is unwilling to pursue within the USA for reasons of civil liberties and politics. If there really was a 'war' on drugs – a goal that could be justified given the terrible social damage caused by them – then it is surprising that known drug-dealers are not summarily slaughtered.

There is no rule that protects democratic societies and Western states from domestic challenges such as narcoter-rorism; indeed, both are increasingly affected by what can be seen as higher levels of potential instability, although this is a long way from the classic civil war. This instability can be viewed in several ways. On one level it can be seen as an inevitable by-product of organized life, in some fashion integral to human society. In other words, the inability to cope with dissent short of violence should not be seen as a failure of incorporation but rather as an inherent characteristic of social life, which is obviously the case in countries such as Afghanistan, but also plays a role elsewhere. This would suggest the need for constant counterpractices and doctrine, not least readiness for militarized policing and for a flexible but firm approach to law and order.

Yet, such a functional approach is overly complacent: in particular, recent, and continuing, social and cultural developments are such that this 'by-product' view underrates the stresses created by violence and the threat of violence, and also the extent to which this violence and related disorders are growing in scale. The first point focuses on the sensitivity of modern society to disruption, seen in particular with economic activity and psychological strain. The economic disruption can be costed, and this costing can be countered by insurance, but current developments, in particular in Russia but also far more widely, suggest that the interaction of criminality and business is such as to distort, if not dominate, economic patterns, and to wrest surveillance and control of much of the economy from the state. Widespread fraud and tax evasion can be seen as facets of the latter.

This situation is not only true of some of the ex-communist states seeking to create a new public culture. Rampant fraud, a lack of any sense of equity in government and damaging governmental responses also characterize a wide range of countries, including, for example, Nigeria and Pakistan. The problem is sufficiently widespread to encourage not just distrust from international economic agencies but also a domestic crisis of government legitimacy. The latter leads to periodic military takeovers, as in much of Latin America, Nigeria, Pakistan and Turkey, but, in turn, also to disillusionment with the armed forces. In such a situation, if the military is heavily compromised by fraud, it appears little different to the mercenary forces that have played a role in the politics of a number of countries in recent years, including New Guinea and Sierra Leone, in the sense that the country is open for exploitation and part, if not much, of its politics becomes the terms on which exploitation takes place. The activities of the military become a serious protection cost for society. This is a way to approach the recent history of some countries, for example Indonesia, and is very much a link between the military and disorder, although war is tangential to the process.

In some states there is a sense that both civilian and military governments are discredited systems of exaction. For example, a lack of legitimacy in both civilian government and military opposition was seen in Paraguay in 2000, when an unpopular civilian government under an unelected president, Luis Macchi, presiding over 16 per cent unemployment, as well as the widespread poverty and corruption that characterizes so much of Latin America, was faced by a military coup by supporters of a former army chief, Lino Oviedo. The coup failed when the rebels failed to win sufficient military support, but, alongside the military coup in Ecuador in January 2000, it was a warning about the readiness to use force to remove apparently unsuccessful and discredited civilian governments, and thus a challenge to those that introduced unpopular economic changes.

Civil conflict can take many forms, and an approach that emphasizes the destabilizing consequences of fraud, in short an illegal but not necessarily violent resistance to government, may seem a long way from traditional conceptions; instead, it serves as a reminder of the need for dynamic definitions of conflict, resistance and control. To look ahead, another version of the same might occur in the European superstate that is on the drawing-board. In order to meet its pretensions to power and its policies of economic intervention, social welfare and regional assistance, the European Union (EU) is likely to have a high tax burden, both direct and indirect; in some, by then, formerly distinct countries, such as Britain, this will run counter to assumptions and past practices of low taxation. Yet there will be no effective way to express dissent within the existing governmental or political process, and this is likely to legitimate a sense of grievance and to lead to widespread tax avoidance. It is easy to see how such processes will lead to violence. Governmental agencies will act as bailiffs, seizing property, and will mount intrusive surveillance operations, which will lead to violence that could easily be widespread. Furthermore, there may well be opposition between centrally controlled European agencies and their local counterparts and rivals, most obviously between European and 'regional' police forces. Whether the net effect is widespread local violence, or a full-blooded secessionist movement, the resulting situation will be difficult to contain. Thus, civil warfare can be seen as the consequence of the pretensions of state structures that do not enjoy public confidence, such as the EU as currently projected, and this will be particularly the case if popularly elected local governments are to be intimidated.

This is a different approach than that which approaches violence within the West as a consequence of radical insurrectionary movements directed against the democratic state, but is a reminder that the process of presenting the state as a moral absolute, legitimated by democratic processes, is one that can and will be contested from a number of directions. Leaving aside the moral absolute, the very notion of legitimation by democratic processes, whether within a state or a superstate such as the EU, can be challenged by anti-authoritarian discourses and practices that deny the 'tyranny of the majority'; while, conversely, claims to the democratic character of particular processes can also be refuted.

The former point is especially relevant, given some of the prospects that can be outlined for human society. In particular, it is probable that resource issues and environmental pressures will cause acute pressures. These will vary by country, but, as already discussed, there are likely to be competing demands for land, water and subsidies, and also pressures relating to environmental quality, which will probably lead to an increased emphasis on government regulation. In some contexts, such regulation is only effective if it amounts, directly or indirectly, by edict or pricing policy, to prohibition. Thus, concern about energy availability and environmental damage may lead to bans on the ownership or use of motor cars, demands on housing stock to the allocation of rooms deemed spare within individual dwellings, water rationing to the prohibition of garden hoses or dish-washers or daily showers, and so on. However much defended on policy grounds, as a mixture of conservation and fair allocation, and however much supported by political processes, such policies are likely to seem radical and disruptive, and to strike many as unfair, illegal and an abuse.

The resulting challenge to these measures may have, in many states, a political and legal focus, but in others the threshold of acts of collective violence may be reached more rapidly. A degree of regulation that might seem acceptable in wartime will not seem so in peace, however much governments may declare 'war' on nebulous targets such as pollution, inequality, poverty, corruption, crime or drugs, as they are increasingly prone to do. Furthermore, the degree of regulation that was accepted or enforced during the two world wars of the twentieth century is now unacceptable, thanks in large part to the social changes of the late twentieth century. Such a prospectus may strike some readers as overly similar to the ramblings of Montana militiamen, but it is intended to underline the extent to which anti-authoritarianism will have a broad range. As a result, government policies are likely to involve states in widespread policing that may well elicit a violent response. This is less true, however, of some non-Western societies, in large part because of levels of repression, but also due to different practices of social conformity and contrasting histories of politicization.

The USA

Reference to the integrative characteristics of national societies and democratic political systems should not necessarily lead to the assumption that opponents of the government are marginal. Indeed, in the case of democracies, the opponents notionally may be a narrow minority of those who voted (if not, as in the USA in 2000, a narrow majority), and/or a clear and large majority of the adult population, as many of the latter will not have voted, and this challenges concepts of legitimacy. In the UK the absence of proportional representation ensures that highly unpopular legislation is passed by a government elected by only a minority of the electorate.

Irrespective of this, there are major limits to the public acceptance of governmental authority in the West. The importance of the American model has to be noted here, although traditions of violent resistance to governmental authority are present in many democratic states, for example France, while in others, such as Italy, there is widespread evasion of regulations and contempt for the prerogatives and practices of government. Suspicion of a standing (permanent) governmental threat to rights and liberties is central to American public culture, while the notion of consent as an active principle is very much present, as is a belief in the value of weak government. In the field of arms, suspicion of government has led, in the USA, to a resilient emphasis on the right to own and bear them, and to an accompanying practice of gun-ownership and use, by both men and women, unprecedented elsewhere in the Western world, although there are high rates of ownership in some Latin American countries, such as Brazil. In the military sphere, traditionally, there has been an emphasis in America on militia and other volunteer service, which led to the development of the National Guard.

As an element of society capable of state-to-state violence, the personal right to bear arms and the American militia tradition are no longer of consequence. America's neighbours are no longer polities that can be threatened by armed citizenry, as they were in the nineteenth century, when Spanish rule was challenged in West Florida, Mexican authority was overthrown in Texas and Fenian raids into Canada, then a number of British colonies, were supported. Instead, through its federal government, America developed the world's leading military in the twentieth century in order to assert and defend its international position and interests. However, American public culture and politics ensured that there was no comparable attempt to increase the military power of the state within the USA.

The federal government has developed surveillance and coercive agencies capable of using force, for example the Federal Bureau of Investigation (FBI) and the Drug Enforcement Agency. Yet, despite concern at the degree of force used against armed millenarians of the Branch Davidian sect at Waco in 1993, these bodies are essentially reactive and heavily constrained by a powerful independent judicial system, and also by the continued power of other governmental bodies, especially at the state level, that are more responsive to popular expectations.

Looking ahead, it is difficult to see this situation changing. Despite the concern of civil libertarians, terrorist actions in 2001, in which large numbers of Americans were killed, did not lead to the creation of an authoritarian society, nor to any fundamental shift in the nature of internal policing. Alarmist talk of a police-state both in the USA and the UK failed to take sufficient note of the practices, such as summary execution, in authoritarian systems such as China, and the military commissions established at Guantanamo Bay were for use against combatants. The American government may have claimed the right to arrest American citizens, including any in the USA, as enemy combatants, and to imprison them without charge until the war on terrorism is over, but this has not been done on any scale.

The strong influence of American models elsewhere owes a lot to the impact of the media, much of which, especially cinema and television, is dominated by production for the American market (the most affluent in the world) and therefore has to make sense in terms of its suppositions. However, the exportability of the American model in terms of state practice is far less clear: whereas American economic policies will continue to be exported, and will retain the aura of American prosperity and economic growth, it is far less clear that this will be the case in other respects. This is true not only of American gun culture but also of dominant American attitudes to the relationship between individual and group provision of social welfare and personal rights, and thus of the notion of a legal restriction, expressed through the constitution, on state powers and state-directed change.

Problems in Europe

To turn further afield: it is unclear whether the model of capitalist economic growth, state regulation, governmental provision of social welfare and centralized definition and control of individual rights, the practice in most of the Western world outside the USA, will continue to be viable, either in economic, social and/or political terms. If it does not, possibly as a result of the rising costs of social welfare for ageing populations, this is likely to be profoundly disorientating for individuals and groups. This disorientation is likely to lead to violence, both against governmental agencies and structures, and against others seen as benefiting from change or threatening traditional practices (e.g. immigrant labour). In short, a collapse of the viability of the state may well lead to an eruption of violence, while attempts to enhance this viability may also lead to resistance.

In the 1930s the crisis of the capitalist model helped produce a new authoritarianism in the shape of Nazi Germany and, less viciously, other states characterized by populism, corporatism, protectionism and autarky; and this was followed by World War II. At present, this seems less likely, if only because state structures and myths are weaker, and social discipline less pronounced, but it is unclear whether the situation will be altered by another major economic downturn. Widespread unemployment, linked to globalist pressures, significantly led, in the West, in the early 1980s and again in the early 1990s, to the panacea of social welfare, rather than to authoritarian governments, governmental direction of national resources and large-scale protectionism. However, economic difficulties can also be associated with the rise of far-right political parties, for example in modern Austria, Australia, France, Germany and Italy, which adopt an adversarial language, analysis and platform, defining and focusing on enemies within and abroad, especially immigrants. Such attitudes and policies are a threat to civic peace, and, were one of these parties to gain control of a major state, then it is likely that its policies would be confrontational.

Russia

The most likely scenario would be in post-communist Russia. Much of the discussion about Russia as a future military power focuses on the damage done to her military capability by the disruption attendant on, and subsequent to, the fall of the Soviet system. The conclusion drawn is that Russia is no longer able to challenge the West, and attention has accordingly switched to the threats posed by China and by 'rogue states'. This may be an accurate perception at present, but two caveats may be offered. First, defeated or 'failed' powers have had an ability to recover rapidly, for example Rome during the Second Punic War,

England in the 1650s, Britain after 1783, France in 1792–94 and Germany after 1918. It can be argued that the role in modern economies of advanced applied technology has removed any such potential in the modern world, and/or that the depths of Russia's decline tend to the same conclusion, but, over the timespan of a century, this account is less clear, while there is the additional factor of the potential for paradigm shifts in military history (whether or not they are presented as revolutions in military history).

Secondly, even if Russia may not be as militarily potent (in relative terms) as it was at the height of the Cold War, it may become potent in absolute terms, and, irrespective of this, retains the capacity to take an aggressive stance towards its neighbours. Thus, despite the pro-Western policies of the Putin government, the potential for fascistic and populist tendencies within Russia is troubling, and this cannot be greatly lessened by reference to its military weakness. The situation is especially difficult in the Caucasus region and Central Asia because they are also inherently unstable, while Russian power also remains a potential element in any Balkan crisis. Russia has retained a military presence in the independent republics of the Caucasus.

Crisis

To return to the Western world as a whole: the widespread decline in senses of nationhood and communitarianism that is particularly apparent in Western Europe, although far less so in Eastern Europe and the USA, will ensure that in depressions the 'haves', whether affluent, or at least defined as such through retaining their jobs, do not identify with those who are less successful. This will encourage the growth of the so-called 'underclass', while the entry into it of those who formerly had jobs may lead to a higher rate of militancy. The 'underclass' can react to adverse circumstances in different ways: most will try to enter or re-enter productive employment, but some will respond with forms of illegal behaviour, such as drug-dealing, which provides employment and status for many, or with behaviour deemed anti-social, such as prostitution. Others will be encouraged to reject existing norms and institutions in a more organized fashion, and this is likely to become increasingly the case because higher rates of educational access across the world will allow more people to express their sense of alienation in a more coherent fashion. Whereas the student revolts of the 1960s were primarily by middle-class children expressing outrage, not suffering, those in the following century will probably be of students and ex-students demanding action because their prospects do not live up to their expectations. From this perspective, the expansion of university provision in many countries may well be destabilizing, creating milieux within which, or as a result of which, extremism may flourish, thus providing fresh tasks for policing agencies. The role of universities in helping engender demand for change was apparent in Iran in 2003. More generally, student politics provide large numbers of motivated and active young people able and willing to act in the centres of power.

Alongside the socioeconomic crisis of unemployment there may also be a rejection of dominant values by the young that reflects an alienation in which boredom and the desire for change, especially intoxicating change, will play a large role. This alienation can be seen in a positive light, as a response to the potentially stultifying grip of former generations, but there can also be a destructive aspect, which takes the form of revolutionary enthusiasms that contribute to a cycle of violence. Looking to the future, it is possible that in reaction to socioeconomic tensions nihilism will have an appeal to many, which is a dangerous prospect because, if on any scale, it will threaten social stability.

Social mobility is seen by most as a one-way street, and one in which there is a demand for instant gratification rather than a long-term process to which the individual has to contribute greatly. The notion of social mobility through education and opportunity for all is a response to economic changes in which the emphasis is on a skilled workforce; and this is then widely seen as a counterpart to democratic civic politics. Yet it is just as likely that fluid labour markets and economic demands will produce a situation in which the few have skills and services that are highly regarded, very mobile and well-rewarded, while the many have nothing in particular to offer and find their wage-rates eroded by global competition, and within their own countries by mechanization and other processes. In short, the processes and problems of globalism will be internalized, and with destructive consequences.

It is unclear whether this socioeconomic model will corrode civic politics. The few may well reject expropriatory taxation, and use the mobility offered by their skills to flee the jurisdiction of high-taxing states, but, more seriously, the many may well be unable to accept the consequences of downward social mobility and economic returns, especially as their expectations will have been very different: most people believe in social mobility, but in their own case are only prepared to consider upward mobility. There will be a sense of alienation by job-holders and citizens that will pose serious problems for political systems, as well, probably, as leading to protectionism, violence towards immigrants and other aspects of domestic and international tension. In short, the great engine of economic growth that powered Western prosperity and democratic politics after 1945 may not go on being so powerful or having such relatively benign effects as it has done hitherto.

In addition, aside from theoretical points about the character of economic growth, there are also practical points about the impact of demographic growth (not least from rising life-expectancy), and thus falling per capita benefits, as well as about the impact both of resource shortages and of competition from non-Western growth. If growth is linked to democracy, and both to peacefulness in international relations and domestic politics, then this diffusion of economic expansion outside the West should not be a threat – an interpretation actively pressed by supporters of free trade. However, it is not clear that this benign model will describe the strains within Western or non-Western societies over the next century: strains that may contribute to a damaging overall impression of uncertainty, if not in some cases decline, which will be potent whatever the experience of the average individual.

Conclusion

In a democratized society aggregate growth-rates are less important for popular perception than the per capita impact, as well as the numbers affected by downward movements. As already suggested, per capita growth is likely to be under pressure from population growth, especially if problems of resource availability increase the real costs of resource extraction and use faster than the rate of improvements in utilization. Put crudely, and the following is subject to qualification, economic growth will occur and at a high rate in some spheres, but not at an overall rate capable of assuaging socioeconomic demands, fears and expectations. The consequences will impact on weak governmental structures, on systems of public politics that are not attuned to long-term crises of this nature and on societies where support for modernization (or at least for many of its consequences) is limited, while opposition is often bitter. The result will be a search for governmental, collective and individual solutions that exacerbates insecurity, and provides issues and occasions for violence.

3 The New Disorder: International Tensions

> Deterrence never was, and cannot ever be, construed as an appropriate response to every military threat.
>
> (Pascal Boniface, 'France and the Dubious Charms of a Post-Nuclear World')[1]

With this chapter we return to a more conventional account of warfare. In the public perception, wars are fought between armies (and navies and air forces), and these are the forces of states. This chapter will look at the likely cause of future wars; while the next will consider how they will be fought. These are not simply matters of speculation, but are actively debated and planned by the military, although it would be inappropriate for scholars to be restricted to such analyses, not least because the experience of the past suggests that, as the military appreciates, most wars do not conform to war plans, and the same is of course true for battles, campaigns and the use of particular weapons. Nevertheless, attention to such planning is a reminder that the purpose of military capability is as much future contingency as present task.

---

[1] In David Haglund (ed.), *Pondering NATO's Nuclear Options: Gambits for a Post-Westphalian World* [1999].

Resource clashes

Looking to the future, it is possible to seek general causes of dispute, specific clashes between individual states and the absence of an effective system of adjudication. The first, in part, relate to the resource issues discussed in the last chapter: rising global demands will be played out in a world in which both the availability of resources and population pressures vary greatly. Although the most sensible ways to maintain and enhance resources require international cooperation, it is likely that confrontation and conflict will arise from unilateral attempts to exploit or redistribute resources, and the situation will be particularly acute with resources that can flow or move across or under borders, for example water, oil, natural gas and fish.

The limited supply, and in some places near exhaustion, of such resources will pose a major problem. Already acute disputes over control of and access to the water in the river systems of the Brahmaputra, Ganges, Tigris, Euphrates and Jordan will become more serious, and will be joined by disputes over other river systems. Water-extraction upstream, for example in India and Turkey, affects countries downstream, such as Bangladesh and Syria. Some tensions, for example between Canada and the USA (the world's most wasteful user of water), or, more specifically, British Columbia and California, and between Mexico and Texas, and perhaps between Kazakhstan and Uzbekistan over the Aral Sea, will be contained peacefully, as those between India and Pakistan over the Indus have been, but they will still engender stress, while other disputes will entail violence. For example, it is difficult to be optimistic about the situation on the West Bank of the Jordan, where water is both in short supply and unequally distributed between Palestinians and Israeli settlers. The latter use about five times as much per person as the former, in part because of extensive irrigation, particularly of crops for export, whereas many Palestinians lack piped water. An optimistic solution would envisage more provision, by desalination, and less use by changing farming practices; but a pessimist would anticipate no such scenario, foreseeing instead conflict over surface water and aquifers.

In addition, around much of the world the offshore availability of oil and fish will ensure that disputes over borders and territorial waters, for example in the South China Sea and the Persian Gulf, will become more serious. Concern over resources has accentuated interest in boundaries. The pressures of growth, the intensification of regional economies and the globalization of the world economy will continue to lead to a situation in which the search for, and utilization of, resources become ever more widespread and important, which will be exacerbated by the exhaustion of established resources, and the possibilities of profitable exploitation in regions hitherto deemed inaccessible or unprofitable.

This will enforce the division of the world surface, on both land and sea, for resource exploration, as production companies have to know from which state they should acquire rights. Thus the intensification of frontier disputes will be a consequence of rising economic demand. Furthermore, the possibility that resources may be discovered will continue to encourage territorial disputes, any of which may become violent or precipitate other tensions. For example, in 1995 Romania and the Ukraine contested the possession of Serpent Island off the Romanian coast as part of a discussion designed to prepare for a treaty between the two states. The Romanian Foreign Minister told the senate that although the island was not then an asset it might become one, owing to oil and natural gas reserves. Under the 1982 UN Law of the Sea Convention, islands, as well as mainland possessions, have to be taken into account in defining maritime zones, and some, such as the Hawar Islands, contested by force between Bahrain and Qatar in the 1980s, and the Red Sea islands, fought over between Eritrea and Yemen in 1995, have become contentious as a result of the actual or possible prospect of oil. In 2000 a dispute between Guyana and Surinam over offshore oil concessions became violent. The importance of these and other resources will encourage external intervention, and greater American interest in sub-Saharan Africa in 2003 was regarded as stemming in part from the growing importance of its oil supplies and the hope that these might permit a reduction in reliance on Middle Eastern oil.

More generally, defence, aggrandisement and security will focus on preserving, expanding and protecting access to resources, as states struggle to meet the needs of their populations and to defend their places in the global economic system. The 1999 Global Environment Outlook report (GEO 2000) produced by the UN Environment Programme, predicted environmental degradation and population growth, leading to the possibility, in the first quarter of the current century, of 'water wars' over scarce water resources in North Africa and South-West Asia. Interest in mineral resources helped encourage and sustain the intervention of Angola, Namibia, Rwanda, Uganda and Zimbabwe in Congo, and also provided profit for military commanders. Similarly, under Charles Taylor, the president forced to leave in 2003, Liberia backed the rebel RUF in Sierra Leone, in part in order to gain control of diamond deposits, which helped ensure that physical control over them became an important military goal.

Economic conflicts

International rivalry over resources may also involve calls to redistribute wealth at a global level, and, as such, will be a new version of the conflicts launched by revisionist powers (those seeking to change the status quo). This attempted redistribution does not have to take the form of military conflict, but the language of combat is, and will be, employed, with reference to global economic relations, more specifically, for example, trade wars or blockades, while, more generally, the concept of exploitation, and resistance to it, is presented in terms of conflict. Aside from raising the question of how best to define war, this issue serves as a reminder that the economic globalization which is seen by some as a safeguard of peace through interdependency is also unpopular, resisted and subject to serious internal strains.

Moreover, the high rates of volatility and interdependency in the global economy are such that trade wars and fiscal restrictions will indeed be able to do an enormous amount of harm, and quickly. They may indeed be seen as the most effective form of warfare for major states, although interdependency means precisely that. For example, trade or investment bans may harm civilians not responsible for government policy in the country thus blockaded, as well as hitting export industries in the blockading country. Although banning imports may encourage domestic production it can also lead to inefficiencies if that production is more expensive, as well as encouraging a pattern of state intervention in the economy that is inherently inefficient.

The extent to which solvency and credit are the product of a confidence that can be rapidly damaged by hostile international actions suggests that traditional forms of trade wars that focus on the prohibition of the movement and/or consumption of goods may be less effective than those that focus directly on the availability and value of money and financial instruments, especially credit instruments. Wall Street and the IMF can have far more influence than warships offshore. This is particularly so, given the declining role of the manufacturing industry in employment and economic activity certainly as far as many Western countries are concerned, while the need of Third World countries for foreign investment creates particular opportunities for international coercion, as does the widespread requirement for foreign expertise and the possibility of applying pressure as a consequence. Thus, in 2003 Royal Dutch Shell was informed by the USA that its willingness to help develop Iranian oil reserves would have a prejudicial impact on its American interests.

In turn, asymmetrical warfare poses a threat to the economic interests of major powers. This was seen with terrorist attacks, such as those of the IRA on prominent buildings in London's financial sector, for example the Baltic Exchange and Canary Wharf, while the al-Qaeda attacks on the World Trade Center in 1993 and 2001 were presumably intended to strike a blow not only at the appearance of American power but also at its financial system. However, these attacks appear to have rested on the mistaken view that such a system could be overthrown by a strike at the leadership, and thus have misunderstood the nature, and undervalued the strength, of the American economic and political systems, specifically the widely diffused character of American power. In response to this challenge, the American government became, from November 2002, the insurer of last resort in the event of another major attack.

To discuss the future of war in terms of measures against foreign investment may appear strange, but in fact it is a reminder of the degree to which war is not simply about fighting, nor violence the best way to force compliance. It also underlines the degree to which international tensions can be between allies and other states that may wish to pursue hostile acts without the commitment and risk of military conflict. As suggested elsewhere in this study, however (p. 109), it is necessary to note the diversity of warfare. A futuristic vision of conflict focusing on exchange rates and liquidity may be appropriate in the event of clashes between say the USA, Japan and the EU, or may be used by the last to discipline Britain and other recalcitrant members. Such an assessment, however, appears far less appropriate for state-to-state relations between less economically developed states, let alone for civil conflict; although expropriatory taxation and related fiscal measures can be seen as a form of class warfare.

What is not clear is how far economic warfare will be used against 'rogue' states, and with what success. Economic warfare after the first Gulf War harmed Saddam Hussein's regime in Iraq, and greatly lessened the value of the country's oil, but it did not bring him down. The current crop of such states use self-sufficiency, brutal domestic control and in some cases (for example Libya and Myanmar) oil wealth to restrict the effectiveness of such warfare, and it is unclear whether future 'rogue' states will employ the same tactics. However, Libya's need for ready access to international investment encouraged its government in 2003 to seek to shed its 'rogue' state status by accepting liability for the Lockerbie bombing.

There is also the issue of timing. Financial and commercial retribution against states can be swift, but may not have a sufficiently quick impact to change the attitude of autarchic regimes since they will be able to pass on the costs to their subjects and use these in order to rally domestic support for the regime. Thus American commercial and financial sanctions against Cuba have helped Castro by making him appear a nationalist, and by providing an excuse for economic difficulties that in fact stem largely from the serious policy failures of his autocratic regime and its centralist controls.

The problems associated with economic sanctions underline the question of how best to take action against states: an issue in which legality, pragmatism and capability interact. At present, there are international attempts to place clear legal restraints on military and other actions, but their effectiveness depends on the state in question that proposes action, on its openness to international pressure and on the nature of its government. The active role of an independent judiciary, legitimated adversarial politics, free press and articulate public opinion in the USA, the UK and Australia did not ensure that international calls for restraint over Iraq in 2002–03 had a domestic echo sufficiently strong to deter action. In many other states, for example Saddam Hussein's Iraq, domestic political institutions and culture had been brutalized and intimidated, and were far less able to influence government policy.

Resource tensions in the future will interact with already existing disputes, for example over borders, as well as creating new ones. To turn to specifics, while it is likely that water and oil will be the foci of conflict, there will also be serious disputes both over the general terms of trade and over their impact on pricing and availability. Import restrictions will be presented as exporting unemployment, leading to demands for retribution. Transport and access will also remain an issue, especially for landlocked states, such as Ethiopia and Nepal, and may become more so as resource questions become increasingly pressing and trade becomes more important in particular economies. Indeed, disputes over transit through Eritrea helped to cause its conflict with Ethiopia at the close of the 1990s, although personal differences between the leaders, and the prestige of their regimes, were also important. India has used landlocked Nepal's dependence on transit rights in order to put pressure on its government. Conflict in Ivory Coast in 2003 affected the nearby landlocked economies of Burkina Faso, Niger and Mali, while the landlocked Central African Republic was affected by conflict in Congo.

Globalism and greater interdependency will exacerbate as much as lessen tensions. For example, greater prosperity, combined with population growth, will help increase Chinese and Indian dependence on imports, particularly of oil, and thus their sensitivity to the availability and distribution of resources. It is unclear whether the two powers will learn how to operate, balance and advance their interests within the international community, as powers seeking a long-term and stable position must, or whether they will strive for advantage in a fashion that elicits opposing actions and, possibly, war. In short, the nature and objectives of regional hegemony, and its relationship with the global situation, can vary greatly.

Regional hegemonies

This is particularly a problem as far as the near neighbours of regional powers are concerned. One of the major risks of conflict at present, and in the foreseeable future, arises from challenges to the notion of regional hegemony. The challenges come from two sources: first, local opposition and secondly the notion of a liberal universalism which, to its critics, is a cover for Western, more particularly American, hegemony. Thus, powerful states, such as China, India and Russia, expect to dominate their neighbours and do not appreciate opposition to this aspiration. This drive for domination is made more dangerous by the relationship between regional hegemony and control over frontier areas, for each of the major powers cited has serious problems with opposition in frontier areas – especially in Kashmir for India, the northern Caucasus for Russia and Tibet and Xinjiang for China, and these problems are likely to continue. Furthermore, both military intervention and political dynamics have made it difficult for these powers to compromise or back down, further ensuring the intractability of the situation, and often making it impossible to adopt or sustain an attitude towards 'hegemony' or 'control' that does not alienate other powers. Thus, the notion of buffer-zones, and the equivalent in terms of attitudes, is not developed, or is only poorly realized.

For China, the principal military challenges in the future may come not from Western powers but rather, as in the long centuries down to the 1830s, from within China and along its land borders. Separatism in Xinjiang may interact with and encourage Chinese intervention in the Central Asian republics, and Islamic links, real or feared, between domestic and international opponents of China may encourage such intervention. Concern about terrorism and Islamic fundamentalism led China to accept American intervention in Afghanistan in 2001. Alternatively, China may be drawn into conflict between Vietnam and its neighbours, as in 1979, or between India and Pakistan, or into Myanmar. The extent to which such conflict triggers confrontation with the West (and/or Russia) will depend on the degree of flexibility that these powers, especially the USA, show, but tolerance may be very limited, not least because of the pressures of liberal globalism, the notion of geopolitical linkages and the extent to which compromise is not a virtue in US public politics.

At a lesser scale, other would-be local hegemons include Turkey, Syria, Iran, Pakistan, Brazil, South Africa, Nigeria and Uganda. Each has reasons to seek to dominate at least one neighbouring state, and for each such dominance is a matter of more than simply the 'realist' calculation of geopolitical need. Instead, strategic cultures (notions about the alleged inherent relationship between national identity and regional hegemony) that reflect ideas and attitudes spread and encouraged by governing elites play a role, as does the drive to appear populist that affects even (and often especially) the most dictatorial of regimes. Furthermore, the instability of regions such as the Caucasus, Central Asia, the Great Lakes region of Africa, Afghanistan and Lebanon encourage intervention by neighbouring powers in order to cement, or to overthrow, particular situations. In these cases, as also more generally with the relationship between civil conflict and the absence or weakness of democracy, there is the lack of an effective alternative to force in order to legitimate and sustain desirable outcomes. In many areas, the interaction of hegemony and neighbouring weakness will continue, and will help to sustain instability. Apart from the regions already mentioned, we should also include West Africa and Sudan, and, looking ahead, possibly Myanmar and Indonesia. As in Central and West Africa, the resulting chaos will not observe state frontiers, and the consequent export of violence will encourage intervention. Thus, India, China and Thailand may intervene in some fashion in Myanmar, as they each border it.

The USA does not face challenges on this scale: there are questions about the long-term stability of Mexico, while Quebec separatism poses a challenge to Canada, but these are different in type. Nevertheless, concern about Mexican stability helped motivate American support for NAFTA (the North American Free Trade Area). This, however, offers a form of stabilization that is different from military interventionism of the type of Syria in Lebanon. It is, and will be, very difficult for Americans to understand the degree to which a number of regional hegemons feel it necessary to overawe their neighbours, in part in order to maintain internal order and also control over frontier regions despite the fact that America itself has followed a similar policy in Central America and the Caribbean.

Regional overawing reflects an entirely different concept of stability and legitimacy to those of modern liberal Western opinion. More generally, the traditional basis of international stability, namely the respect by states of each other's sovereignty, have been eroded, or at least challenged, by the claims of globalist politics based on international standards of governance, and on the panoptic eye, or at least lens, of the media. This challenge has also been mounted against established practices of regional hegemony and balance of power politics. The traditional account of international relations was in fact largely a Western view, in that the mutual respect of sovereignty is a concept that has less meaning in, for example, Chinese and Islamic thought. Despite the fact that Western concepts were spread both during the age of imperialism and in the period of decolonization, and were expressed in a myriad of treaties, alternatives retained a certain potency, which is currently encouraging interest in Islamic views on international relations.

Interventionalism

Across the world, the current view of most governments towards international relations is based on a national integrity that is expressed in terms of sovereignty, and this is challenged by external attempts to prescribe rules, whatever their source. However motivated by liberal sentiments or defensive strategic goals, interventionist internationalism is likely to cause conflict unless it is restrained by a prudential respect for traditional power politics. It is a difficult balance to strike, not least as the task will fall on American leaders who are neither best prepared for it, nor able to play to a domestic constituency that understands international power politics and the validity of alternative value systems. Standing up for the USA strikes a chord with American public opinion, but other states standing up for themselves, especially if in different terms, do not win American understanding, with the exception of Israel which is seen by an influential section of American opinion almost as an extension of the USA. In addition, American attitudes will be affected by a determination to see as normal an economic world that enables the USA to use its research skills, technological capability, investment capital, operational economic control and purchasing power to gain a very disproportionate share, in aggregate, and even more in per capita terms, of global resources.

In 1991 the Gulf War brought a happy convergence of Western resource interests, in Middle Eastern oil, and a widespread ready response to a clear-cut aggression by an unpopular non-Western power that made it relatively easy to elicit and sustain widespread international support for action. Such a convergence did not recur with Iraq in 2003, and that helped to explain the international bitterness caused by the war. Although the war on terror proclaimed by President George W. Bush in the aftermath of the September 11 attacks required a global range if the networks, bases and support systems of al-Qaeda and related organizations were to be defeated, it is unclear that Iraq was in fact a supporter of al-Qaeda. In contrast, there was more international support for action in Afghanistan from 2001, as the Taliban regime clearly did support al-Qaeda, and the notion, based on the experience of Afghanistan, that 'failed states' provided a base and cover for terrorism encouraged such interventionism.

In the case of humanitarian rationales for intervention, the issue of mission-creep and the question of national interests were raised when France intervened in Congo in 2003, sending troops to reinforce UN peacekeepers, and, the same year, when the USA considered intervening in Liberia, but instead restricted itself to providing logistical support for Nigerian peacekeepers. It will remain unclear whether interventionism is based on a sensible assessment of means and goals. On one level, such an assessment will depend on the restraint displayed in the cartography of Western concern. It is highly unlikely that there will be a war to drive China from Tibet, or military intervention to try to resolve conflicts in the Caucasus or central Africa, and it is likely that the reality of a new world order, whatever its rationale and goal, for example a worldwide attack on terror, will be more restrained than the language.

That analysis can suggest a misleading complacency, for the notion of a clear or static boundary defining a zone where it is safe to intervene is questionable, as also is the notion that other states will readily be able to distinguish a language of global order from the practice of a more limited pattern of commitment. Wars frequently arise as the product of the interaction of bellicosity and what could be regarded as a misjudgement of the resolve of other powers; and there is no reason to believe that this situation will alter, for distrust is a constant factor, as is the multitude of reasons why powers misunderstand others. The classification of terrorist groups provides an instance of this problem. Western abhorrence of terrorism, a global statement, clashed not only with past willingness to support terrorist movements against states deemed unacceptable, for example in Afghanistan and Nicaragua in the 1980s, but also raised the issue of how best to regard contemporary movements opposed to such states, for example the Mujahideen formerly armed by Iraq against Iran, and now a possible ally for the USA.

American attitudes

Secondly, liberal internationalism will encounter the problem that the major Western power, the USA, will continue to seek to be a global hegemon, rather than an internationalist power, and will continue to have an ambivalent relationship with the constraints of collective security, especially with the UN. The USA is particularly unhappy with the notion of UN direction of operations involving American forces, and the possibility of UN influence, and still more control over, international policy towards, say, Israel or Cuba. More generally, it is unwilling to accept the subordination of American juris-dictions and interests to their international counterparts.

These positions are not without point. To take the first, it is far from clear that UN military structures are adequate: there are particular problems with UN staff, organization and practices, and the use of the UN presup-poses acceptance of the legitimacy of the sovereign gov-ernments represented in it. While this may be the case in international law, it is easy to appreciate why Americans (and others) who live in democratic societies and have democratically elected governments will continue to oppose giving authority, let alone power, via the UN to the representatives of undemocratic societies and autocratic, indeed frequently vicious, regimes. Both are well repre-sented in the UN, which, despite its aspirations, reflects the lack of any homogeneity of culture and values in the world. To turn to the future, it is also easy to predict a rift, which will undermine the UN, between international bodies using egalitarian arguments to demand the redis-tribution of resources and Western societies resisting such processes.

Aside from their difficulties with the UN, American politicians and public opinion will continue instinctively to think in unilateral, not multilateral, terms, and import-ant tensions will remain between the external constraints that alliance policy-making entails and the nature of political culture in the USA, which tends to be hostile to compromise with foreign powers and also not to trust them to observe their commitments. The notion of American exceptionalism does not encourage the exigencies of compromise, while the history of American foreign policy is that of American leadership, and thus does not lead to an interpretation of alliances as based on mutual needs. Indeed, American politicians argue that Congress can supersede treaty obligations.

It is unclear how far American politicians will remain convinced of the value of longstanding commitments, for example in Europe, while, despite the interest shown by President Bush in 2003, the history of American engagement in Africa suggests that it is entirely possible for the USA to follow a spasmodic course, dictated by immediate concerns and, more generally, by neglect. But for the alliance structure provided by NATO, it is possible that that would also be the pattern in a post-Cold War Europe. Conversely, American concerns about the Islamic world and China might lead future American governments to seek European and Russian support, but what is less clear is what they will choose to do if it is not possible to have both, let alone what they will do (other than express anger) if support is not offered on acceptable terms.

America and the problem of allies

For the Americans, European enlargement is a goal, as it can lead to the 'anchoring' of strategically important states, particularly Turkey and even Russia, in the Western camp, but this process may not work from the American perspective. A stronger Europe may be neutralist or, as was shown by France, Germany and Russia in 2003, critical of the USA, forcing the Americans to look elsewhere for allies. This is particularly likely to be the case if Russia plays a role in European security structures, which may well lead to the neutralization of NATO, directly, or through its influence on European powers such as Germany. Conversely, a future Europe may exclude, and be in rivalry with, Russia, giving America a choice of allies against China; or the EU may fail, leading to a situation in which 'Europe' involves a variety of options and problems for the USA. As a parallel to European developments, Turkish policy in 2003 indicated that a stronger Turkey might not follow American policy in the Middle East.

In the event of tension between China and the USA, it is far from inevitable that the latter can rely on Russian support. Although each historical example has its particular characteristics, it is worth noting that in World War II, despite its Manchurian empire and ideological hostility to communism, Japan turned against the USA, not the Soviet Union. Nevertheless, a Russia determined to retain its position in the Far East, needing to secure its Siberian resources and wary of Chinese–Islamic links, for example cooperation with Iran and Chinese interest in Central Asian resources, is likely to see China as hostile. On geostrategic lines, this should encourage Russian cooperation with the USA, but such a relationship can easily be mishandled in an atmosphere of distrust, and as a result of the search for national interest. Concern over American globalist pretensions and particular policies has indeed led to an identification of the USA as a threat that has encouraged a positive Russian response to Chinese approaches, which has been clearly revealed by the large-scale arms sales to China that have helped keep the Russian arms industry buoyant. Furthermore, in an echo of the dispute over responsibility for the appeasement of Nazi Germany by the Western powers and the Soviet Union, a less than robust American response to Chinese intimidation of Taiwan may lead the Russians to feel that America cannot be trusted, while concern about the position in the Russian Far East may also lead to a sense that it would be better for Russia if China and the USA clashed.

All of these factors will create problems for American governments, which will accentuate the habitual American distrust of allies and lend new force to it, because hitherto the Americans have not had to anchor their foreign policy on cooperation with a state and society deemed hostile. Nixon's reconciliation with China was important, but China was not America's key ally. Yet, it is unlikely that America will wish to confront China without at least a powerful ally or allies. The Americans will certainly face difficulties if this ally is Russia, but for a number of reasons it is unlikely that Russia will make a rapid recovery from her current problems. The demographic situation in Russia, in which male life-expectancy fell from 64.2 in 1989 to 60.9 in 1997, will remain harsh, affecting the numbers in healthy, productive employment, and there will continue to be major difficulties in obtaining financial stabilization and economic growth. The development of close relations between criminal groups, politics and business is a threat to Russian stability, and in particular to the ability of the government to gain sufficient revenue from taxation, which will challenge attempts to use resources to any planned end, including military modernization. As far as Russian geostrategy is concerned, it is likely that Russian governments will find it difficult to sustain a high level of military preparedness, let alone capability enhancement, despite the hopes of their generals. Furthermore, Russia will remain challenged by the situation in Central Asia and the Caucasus, and by uncertainties about developments in the Islamic world. This will result in Russia having a less benign strategic situation nearby than the USA, whatever might happen in Mexico or Quebec. The problem will be compounded by Russian assertiveness, and by the possibility that such assertiveness will win domestic political support, as with Putin's war with Chechnya, which helped him win the Russian presidential election in March 2000, but subsequently made it difficult to negotiate a settlement.

These problems will not prevent the USA from considering using Russia against an aggressive China, as the two states have a long frontier and Russia retains a significant military capability. The difficulties facing Russia, however, will prompt stronger US interest in alternative allies, the most obvious of which are India and Japan. Both are major military powers, and each feels challenged by another state that is unpredictable and that may be supported by China: Pakistan and North Korea respectively, while the Japanese economy will probably remain well integrated with that of the USA, and that of India will become more so, as the state socialism of the past is abandoned.

Yet, neither India nor Japan are comfortable in the role of ally against China. Both feel vulnerable, and rightly so, and neither wishes to lose the degree of flexibility about policy that they currently possess. In addition, despite considerable investment in military modernization, their force structures and doctrines are not designed for an offensive war against China. If, in the event of war between China and the USA, India and Japan simply protect their space against Chinese attack that will be of only limited value to the Americans. The failure of the much subsidized Pakistani and Turkish militaries to come to the support of the USA in the Gulf War of 1990–91, and the refusal of Turkey to permit use of its territory for operations against Iraq in 2003, are warnings about any reliance on India and Japan. For political reasons, they are unlikely to take part in operations against China, and, indeed, in the event of such a war, the Indian military is likely to be most concerned about Pakistan, which has aligned with China against India.

Other allies will probably be found wanting in any clash with China, for states as varied as Australia, Israel and Germany will not meet the requirements of US policy. Indeed, past examples, such as the Vietnam and two Gulf wars, reveal a conditionality in support that is unwelcome to America; while, on the other hand, allies feel a lack of consultation and a concern about American policy-making processes. Concern about consultation is understandable in political terms, but poses difficulties for military planning in both the pre-conflict and conflict stages of any confrontation or war: both pre-conflict and conflict are now very high tempo, while alliance co-operation raises issues of security as well as speed, as was seen in the Kosovo war of 1999.

Yet, the consequence of democratization is a demand for accountability in government that makes it difficult for states to accept the leadership of another or, indeed, the consequences of membership in an alliance. The notion that the latter may lead to unwelcome steps is antipathetical to the democratizing principles and practices already referred to, which reflects the extent to which nationhood does not act as a building-block for global cooperation, but rather both as a delimitation of concern and a demand for independence. This is likely to be a key issue throughout this century, and one which will provide a background to military operations. Nationhood in many cases will also be an expression of atavistic tendencies, which will not preclude alliance politics but will condition their character.

Aside from this general point, there will also be particular tensions and dynamics within the American-led West as American leadership acts to contain the ambitions of other powers. In Europe, there is a serious problem with France, where anti-American sentiment plays a major role in policy-making, and, albeit to a lesser extent, with Germany, the most populous and economically powerful state, and one that is very close in policy terms to France. At present, German public culture is resolutely against territorial expansionism, but that is not the same as abandoning the desire to use national strength to secure and advance interests, as was shown clearly with the collapse of Yugoslavia when Chancellor Kohl's support for Croatia, despite its rule over Serb-inhabited areas and the anti-democratic nature of Croatia's government, helped to precipitate large-scale conflict between Croatia and Serbia. Looking to the future, the area to the east and in particular south-east of Germany appears, at least in part, unstable, and taking the notion that power abhors a vacuum, and that it is difficult to contain problems (as well as following business investment), German politicians may both want and feel it necessary to take the lead. Irrespective of their intentions, this may provoke local opposition and the hostility or fears of other states, particularly Russia. This concern will lead some to see American power as a guarantee, and potential curb, of German intentions. This tension played a clear role in the critical response of what Donald Rumsfeld termed 'New Europe' to the Franco-German attempt to lead the EU in 2002–3 in opposition to America's Iraq policy.

Enforcing a world order

More generally, globalism and worldwide commitments will continue to pose major problems of prioritization for the West, which may encourage restraint or, alternatively, a resort to war in order to try to settle an issue. The multiplicity of problems and commitments facing Western powers will limit the ability to respond to fresh problems: an issue dramatized for the USA by the need to plan for 'two-front' commitments. This may lead to the reliance on a rapid response in a particular crisis – the use of force in order to end the need to use force – which risks proving as unsuccessful in the future as it has often done in the past. The 'high-tech' nature of the militaries of leading powers is also important, as they are designed for a very forceful response. For example, as in Chechnya, the Russian military will continue to show a preference for firepower, and, more generally, the dominance in doctrine and practice of Cold War concepts and training, rather than those possibly better attuned to low-intensity and counterinsurgency warfare. Although far more balanced and multipurpose in structure, training and doctrine than the Russians, and understanding the need to use this variety in practice, the Americans will continue to have an emphasis on force and response doctrines and tactics, as well as on firepower. While understandably designed to minimize the exposure of ground troops to risk, such doctrine and tactics offer little of the alternative of militarized neighbourhood policing, with its stress on cooperation with local communities, but in many contexts it is the latter that is required, not only politically but also militarily.

Irrespective of the military means, long-range as well as universalist conceptions of American interests will interact in a positive or contradictory fashion with the global pretensions of the UN, and the sense that the settlement of international problems is a responsibility for neighbouring and other powers. The resulting premium on order, of a certain type, will encourage concern in and about distant areas, and, conversely, insecurity in these areas about the prospect of outside interventionism. This order will also be difficult to enforce. Thus one consequence of globalization, the wider implications of local struggles, will continue to be seriously destabilizing.

Weapons of mass destruction

Concern about distant areas, however, will not simply reflect universalist aspirations. Instead, both military and political pressures will encourage concern and confrontation with, and even intervention in, other states, including those at a considerable range. To that extent technology has transformed geopolitics, and hence military goals, capability, preparation and planning. The range of readily available modern weaponry is such that it is difficult to ignore hostile developments elsewhere, specifically the spread of missiles and the production of weapons of mass destruction. Near-America ceases to be Cuba – which is, indeed near – and becomes North Korea; while Rome can be attacked from Baghdad as well as Benghazi. The pace of change in this field will continue to be rapid, not least because it is essentially a case of the diffusion of tested technology, which can be used for the dispatch of weapons of mass destruction, rather than the invention of 'super' weapons; as a result, concern will increase, as will demands for action.

One of the most important developments in forthcoming decades will not be the use of new military technologies (which tend to dominate the futurology of war), but the spread of established technologies to states that hitherto had not possessed them. Thus, for example, second- or third-rank powers, and also non-state movements, might be tempted to use nuclear, chemical and bacteriological weapons of mass destruction, not least because they are not inhibited by the destructiveness of such weaponry, which is also relatively inexpensive. The nature of international asymmetrical warfare, therefore, may become much more threatening to the more 'advanced' power, and its home base may be subject to attack: the concern that affected the USA when letters with anthrax spores were received in 2001. In turn, this vulnerability will help to drive investment in new technology, just as it will encourage the maintenance and upgrading of a strategic nuclear arsenal and the relevant delivery and control systems, for deterrent purposes, by the USA and, possibly, by other powers, although such a policy is expensive.

Despite the sarin attack on the Tokyo Underground in 1995, the prospect of bioterrorism was not at the forefront of defence planning or public concern in the 1990s, but the situation radically changed in 2001, leading to immediate responses, including the stockpiling of smallpox vaccine and efforts to control the production, trade and storage of dangerous chemicals and pathogens. Research in the understanding and detection of agents was also stepped up. Thus, the American Defense Advanced Research Projects Agency sponsored research in hyper-spectral imaging in order to permit the detection of chemical and biological weapon manufacture from a distance, which helped address one of the major problems with surveillance.

The extent to which the spread of nuclear, chemical and bacteriological weaponry will encourage the prophylactic use of force – wars, or at least the use of force, to stop the deployment of weapons systems – is unclear. In 1981 the Israelis bombed the Iraqi nuclear plant at Osirak (using American-supplied aircraft), claiming that the Iraqis were manufacturing nuclear weapons. Similarly, the need to pre-empt possible Iraqi use of weapons of mass destruction was cited by the USA and the UK as a reason for the attack on Iraq in 2003. However, there has been no other comparable use of force. India and Pakistan responded to the development of nuclear weaponry by the other by stepping up their own production, rather than by launching pre-emptive strikes. The Israelis have threatened such strikes against Arab powers and Iran, but their wish to avoid war may instead encourage caution, and the Americans are unlikely to countenance the use of planes they have supplied for such purposes. Similarly, the Americans and the Europeans have chosen to respond to the development of long-range weaponry by North Korea and Libya by defensive schemes and attempts to restrict the flow of technology, rather than by offensive action. Thus in 2003, when North Korea admitted to developing its nuclear capability, the USA sought a peaceful solution.

The problem posed, however, related not only to North Korean capability but also to its willingness to export missile parts, both in order to obtain money and to gain missile technology in return.

It is possible that this policy of relying on deterrence rather than pre-emptive attacks will continue, but it does face serious problems, including the nature of the defensive technology, which is untested in war and reliant on very high degrees of accuracy, with scant room for error. In particular, anti-missile weaponry systems can be confused by decoys, which are becoming increasingly sophisticated. Furthermore, relying on deterrence risks blackmail by governments that are regarded as unpredictable and unlikely to be restrained by the prospect of 'mutually assured destruction'. The 'Would we have fought Hitler had he had the bomb?' question invites the response 'What alternative was there?'; but this may encourage pre-emptive strikes in the future, both by states that feel unsure of their position, and by the USA in its own interests and in its position as leader of Western internationalism, however much that position is questioned by some European states, especially France. American weakness when/if it occurs may also lead a fearful Israel to launch pre-emptive attacks.

This problem was highlighted in 2003 when, from the perspective of nuclear proliferation, the USA went to war with the wrong Islamic state: Iraq, not Iran. Using Russian technology, Iran, it became clear, had developed its programme to establish a nuclear reactor to include a uranium-enrichment plant at Natanz and a plant to produce heavy water at Arak. As a result, the peaceful claims made by the Iranians were doubted, and the prospect of limiting nuclear proliferation in the region called into question. In turn, Iran felt vindicated in its nuclear plans by its inclusion in Bush's 'axis of evil'. Similarly, in June 2003 the North Korean official news agency declared that the state had a right to develop 'nuclear deterrence' against aggressive American plans.

Whether or not opposing states are armed with weapons of mass destruction, the USA and the UN will continue to be faced by the problem of eliciting consent to their views and their conceptions of international order. Both will also face the problems of enforcing a verdict and, more generally, of the wasting quality of international order, in other words of the inherent fragility of such order in the face of changes and demands. The international order will require consent for its maintenance, rather than simply force, because force in the absence of consent may work for blasting aliens, but is of limited value when dealing with humans. This is not intended as an anti-war remark, because the nature of communities and international systems is such that it will not be possible to rely on consent alone; however, to move to the opposite position and assume that in the future force can operate without an attempt to build up consent is of limited value. Indeed, the experience of the last 150 years suggests that those brought low by force have an ability to reverse the verdict, either peacefully or by violence. The high cost, political, economic and financial, both of future war and of the far more lengthy process of postwar peace maintenance – an issue facing the USA in Iraq in 2003–4 – will be such that a verdict that has to be maintained by long-term vigilance and the periodic use of force, the prospect held out after both the al-Qaeda attacks and American interventions in Afghanistan and Iraq, will seem unacceptable to most states and societies. However, this might encourage authoritarian and intolerant states to resort not to the search for consent but to harsh domestic and international tactics, such as 'ethnic cleansing' and the encouragement of instability in neighbouring and revisionist states. In turn, force, not consent, will be the necessary response to such tactics.

In the recent past, the most successful way to avoid the future unpicking of the verdict of war was the attempt to rebuild a civil society to which authority could be entrusted. This was seen in the treatment of Germany, Japan and Austria after World War II, as well, more generally, as the American attempt to stabilize Europe through the Marshall Plan. However, such sociopolitical engineering was far less successful when attempted by the Americans in South Vietnam, the Soviet Union in Afghanistan and the Israelis in Lebanon. This suggests that such policies work best when there is at least a degree of commonality in socioeconomic and politico-ideological circumstances, a process that encouraged the use of Turkish forces for pacification operations in Afghanistan. Furthermore, the policy pursued in Germany, Japan and Austria required a total victory, the formal surrender of the defeated regime and a massive postwar commitment by the victors, including occupation by large conscript armies, none of which is very common in modern warfare. The absence of such a surrender was a serious problem in Iraq in 2003.

In the future, it is far from clear that such political costs from war and postwar are bearable; they will also exceed the available military resources of even the most powerful state. The alternative will be a negotiated settlement, rather than total victory, which, is in fact far less typical as an attainable objective than the rhetoric of politicians, the experience of World War II or the accounts of science fiction might suggest; and this is likely to remain the case, so that most wars will continue to end in negotiated settlements, not capitulations.

Yet it is unclear how a future war would lead to such a peace. To do so, it would be necessary to wage a war in which the opposing society and government was not demonized, but that would be difficult, given the need to offer a public justification for conflict, and given the unwillingness of much of the public to accept anything less than the Holocaust justification. 'Serbs are like Nazis and therefore we must act' was the gist of the argument pursued by the West in the Kosovo crisis of 1999. Despite the dangerous viciousness of the Milosevic regime it was an argument that underrated the complexities of the situation, and represented a worrying inability to learn from the earlier Bosnian conflict when the Croats pursued 'ethnic cleansing' with as much zeal as the Serbs.

Such demonization makes it difficult to negotiate peace. In the conflicts between Western powers and both Iraq and Serbia in the 1990s and Iraq in 2003, the lack of an adequate exit strategy from the war became a serious cause of post-conflict political problems, and thus led to a continued need for military deployment. It is likely that similar problems will affect future Western interventions. There will certainly be no international agency capable of offering an effective system of adjudication, nor a global army that would lessen some of the current problems of internationalism. The divergence in views among the member states towards, first, the objectives of government, secondly, the nature and value of human rights and, thirdly, the use of force, will continue to be a challenge to the operation of UN forces and to the ability to agree goals.

For a long time these problems were in part masked by Cold War divisions, but it is now clear that they are of a more lasting character than the struggle between communism and the 'Free World'. After 1945, it had been hoped that the UN would control standing forces, and the Military Staff Committee that advised the Security Council sought to agree the allocation of units from its five permanent members. However, they were unable to agree and abandoned the task in 1948. As a consequence, the UN developed a system that is very likely to continue in the future: a resort to *ad hoc* forces for its operations. Such combinations of the 'willing' have also been the pattern for armed actions by other collective security systems, for example by the Warsaw Pact[2] powers in Czechoslovakia in 1968 and by NATO powers in the former Yugoslavia in the 1990s.

More generally, aspirations towards global cooperation fell victim to the central role of individual states, the reality of the Cold War and their own impracticability. These aspirations, nevertheless, continue to be important to the terminology of UN operations, and they encourage an emphasis on 'peacemaking' or 'peacekeeping', not 'war', which suggests an ability to control the commitment, and a limitation of effort that does not disrupt peace, or peacetime expectations of the relations between individuals and the state. Such terminology and assumptions could, however, be misleading.

Assumptions, nevertheless, are important in influencing the parameters within which decisions are taken. These indicate the delegitimation not of war but of some forms of war. The notion that force should not be

[2] The military association of most communist powers.

employed in order to alter frontiers was included in the UN Charter and reaffirmed in a UN resolution of 1970. Aggressive war was also prohibited in the Charter. Together with the idea of the sovereignty of the state, such notions have been important. They have not ended war or aggression, but they have changed their character. For example, seeking to overthrow a hostile government by supporting an insurrectionary movement is now far more effective than formally waging war. This is true even if there is a vast disproportion in the resources of the two states, such is the attitude towards aggressive warfare.

Such aggression was punished by international action when Iraq invaded Kuwait in 1990. The widespread nature of the coalition forces against Iraq was impressive, but so also was the overwhelming refusal of most of the world's states to recognize the Iraqi annexation. In part this was a consequence of America's leadership of the anti-Iraqi coalition; and it was no accident that the small number of states that recognized the annexation included such opponents of the USA as Cuba and Sudan. However, far more than this was involved, for there was a wide-spread feeling that in seeking to extinguish a state Iraq had used violence in order to challenge the entire inter-national system, that, in short, the pursuit of such politics by war was unacceptable. The American-led recourse was war, not blockade or any other expedient, but war under the mandate of the Security Council. The very different context of the conflict in 2003 in part reflected the view that however vicious, indeed potentially dangerous, its government, Iraq was not the aggressive power in that crisis.

China versus the USA

'Peacemaking' or 'peacekeeping' would scarcely describe a clash between the USA and China, even if the cause of the conflict was an American response to Chinese military pressure on Taiwan, or indeed a Chinese response to American action to prevent an enhancement of the North Korean arsenal of advanced long-range weapons, both of which are feasible. The Chinese showed with their intervention in Korea in 1950–53, their conflict with India in 1962 and their invasion of Vietnam in 1979 that they were determined to assert their power when it seemed both necessary and possible. China's prosperity has helped to make it more assertive, as has its sensitivity to criticism of its domestic situation, and the sense of the need to return to an intrinsic great power status that was compromised by the West in the nineteenth century. More specifically, Russian weakness has both made China less vulnerable and decreased its interest in good political relations with the USA, at the same time as economic links have grown much stronger. Russian weakness has also been bad news for the Western alliance because it has encouraged neutralism in Europe. In particular, both French and German politicians continue to believe that it will be possible to manage Russia through some sort of partnership, and that this partnership reduces the need for cooperation with the USA, which is likely to be a major dividing point within both the EU and NATO.

Russian weakness, and the absence of a universalist Russian ideology comparable to the liberal global interventionism of the USA, has made it most likely that China's opponent in any great power struggle will be the USA, which will also reflect America's specific interests in the West Pacific and, to a lesser extent, South-East Asia, and a more general concern to preserve the international order that it has created. In the case of communist China, there is also a tradition of American hostility, alongside that of a search for better relations. This tradition, and its location in terms of the political culture of the two states, provides both sides with a ready vocabulary for dispute. The 'comprehensive engagement' that America under President Clinton sought with China was designed to limit Chinese revisionism of the international order, and included unsuccessful attempts to dissuade the Chinese from transferring advanced weaponry to hostile states. Under President George W. Bush, such engagement was constrained by greater American criticism of what were seen as hostile Chinese policies; although the wish to secure Chinese acceptance of American policy towards first Afghanistan and subsequently Iraq helped temper earlier tension.

China signed the Nuclear Non-Proliferation Treaty in 1992, but has only episodically observed its provisions. It provided Pakistan with parts used in uranium enrichment, M-11 missiles capable of carrying nuclear warheads and information on how to build such warheads. Indeed, China has shown a particular interest in advanced military links in South Asia, from which Russia and India, both potential opponents, can be challenged. Thus in 1991, Iran was provided with 1.8 tonnes of uranium, a useful raw material for the nuclear industry. China and the USA have clashed over these transfers and over American missile defence plans, and these disputes are likely to get worse. In 2000 the American government was warned that plans for a comprehensive missile defence system might lead the Chinese to increase dramatically their number of warheads in order to be able to overcome whatever system the Americans could deploy. To this end, China plans to upgrade its intercontinental ballistic missiles, making the missiles more mobile and the warheads more accurate. These missiles will be able to reach the continental USA. The entry of India and Pakistan into the ranks of the nuclear powers has made China's position look less secure and has increased the sensitivity of the issue of missile defence and deployment. Chinese interest in developing intermediate-range missile systems are in part a reply. Any increase in Chinese nuclear weaponry is likely, in turn, to lead to a response by India and Pakistan.

China's long-term Versailles complex – the sense that it has been wronged by history – remains powerful, and needs to be considered when reassurance is offered about the intentions of the current generation of communist leaders. The return of Hong Kong and Macao removed prime irritants, but foreign concern about human rights within China, especially those of religious groups and Tibetans, is seen as unacceptable. Furthermore, the very fact of American power, let alone the ability of America to act to defend Taiwanese sovereignty, are unacceptable to Chinese policy-makers, and may become more so in the future. This parallel with French attitudes after the Congress of Vienna in 1814–15, and German revisionism after the Peace of Versailles of 1919, serves as a powerful warning that the idea that the status quo should be the basis of future relations is inherently unacceptable to certain powers, as it is perceived by them as both cause and symbol of international instability. Revisionist powers are generally seen as 'rogue' states when they seek to alter the situation by means outside those of peaceful negotiation, but, being realistic, such negotiation rarely serves the cause of revisionism, and to pretend otherwise would be complacent. As a consequence, revisionism is inherently a cause of instability, as is the response to it.

Aside from differences in South-East and East Asia, it is possible that China and the USA will clash in the future as a consequence of growing instability in the Pacific, where economic tensions in many of the island groups have exacerbated, and been accentuated by, ethnic tensions. In 2000 alone, this led to violent attempted coups in Fiji and the Solomon Islands. Since 1945, the Pacific has not been an international battlefield, as the defeat of Japan in World War II was followed by an American hegemony that was enhanced by the support of Australia and a rearmed Japan. It is possible, however, that, as China becomes a stronger maritime state it will seek to challenge American interests. Whether successful or not, this will lead to a degree of instability in the Pacific, and will further threaten relations between the two states.

It is also possible that Chinese membership, alongside Japan, South Korea and ASEAN (the Association of South-East Asian Nations), in the ASEAN+3 network may temper Sino–US tensions, as many fellow-members have close ties with the West, but in the long term if global economic tensions become more serious it may also provide China with a degree of regional support. China indeed has been trying to persuade Asian powers such as Malaysia and Indonesia that their economic interests are best served by looking to it, rather than to the USA or Japan. Although capitalism (like democracy) has been seen as a force for peace, capitalist powers have fought each other, as in World War I, and, whatever their shared interests through trade, it is far from clear that a capitalist China will be a more welcome partner for a capitalist USA than communist China was. Indeed, a free-market China may find it easier to capitalize on East Asian disquiet with Western economic policies and financial hegemony, disquiet that grew after the 1997–98 financial crisis, because the feeling that the West, through institutions such as the IMF (rather than incompetent, if not corrupt, East Asian fiscal policies) was responsible for the crisis is deeply ingrained, and has encouraged an East Asian search for economic autonomy. American and European protectionism and/or limited economic growth will accentuate these tensions. The notion of a regional pact between China and Japan may appear incredible, but the EU has brought France and Germany close together, and may do the same for Germany and Poland. However, China's ability to act as a regional leader will be undermined by continued suspicion of its intentions, especially by Japan and South Korea.

A conflict would test the military effectiveness not only of the USA and China, but also of their allies. It is far from clear that such a conflict would be settled by a technological lead; although it is also worth noting that, while such a remark would generally be seen as a warning against the inevitability of American victory, the Chinese themselves have made advances in their military capability, in part from borrowing American capability, by trade, purchase and espionage. The relative ease of doing so serves as a reminder of the difficulty of maintaining a lead in technology, provided the industrial infrastructure to permit production is present in both countries. Indeed, the continued spread of advanced engineering and electronics will ensure that in the future the Western dominance of advanced weaponry and the relevant control systems will be increasingly challenged; although, in turn,

American investment in research on such weaponry and systems is unmatched, and the weakness of China's military-industrial base is such that recent expenditure on the military and on this base have not remedied their major weaknesses *vis à vis* the USA. The Chinese need to purchase advanced weaponry from Russia, which, so far, includes guided-missile destroyers, submarines and modern fighters, is an appropriate indicator, while the display of American capability in conflicts from 1991 have indicated China's relative weakness.

If advanced weaponry was used by America in an all-out conflict with China, it is not clear whether it could fulfil objectives or survive a rapid depletion rate. As in the Vietnam War, the timetable of conflict by the two sides might be very different, and such that the short-term high-intensity use of advanced technology by the Americans achieves devastating results, but without destroying the Chinese military system or overcoming the political determination to refuse American terms. In addition, the experience of recent conflicts, especially that with Serbia in 1999, suggests that 'just in time' procurement systems are inadequate, as they provide neither sufficient weaponry nor the sense of confidence in reserves that is necessary for operational choice and for planning. This inadequacy is a product both of a financial stringency that mirrors constraints over manpower availability and the modernization of military technology, which ensures high rates of obsolescence, and thus also discourages stockpiling. In particular, the lack of adequate weapons stocks has been a problem that has affected the use of cruise missiles.

Islam

A similar challenge will affect the USA in the event of any clash with a major Islamic power, such as Iran, or Egypt were it to be taken over by fundamentalists. The Gulf wars of 1991 and 2003 are only a partial guide to the problems that might arise, because Iraq was isolated, the USA was aligned with Islamic states, especially in 1991, the Iraqi regime lacked mass support and the international situation elsewhere permitted a massive American military commitment. More generally, the military challenge posed by Islamic assertiveness, whether fundamentalist or not, has been underrated because there is no single major Islamic state that enjoys power comparable to say India, let alone China or Russia. Furthermore, powerful rifts within Islam, not least between Shias and Sunnis, and between secularists and fundamentalists, are likely to continue, and may well become more savage.

Nevertheless, it is anticipated that by 2030 Muslims will make up about 30 per cent of the world's population. Furthermore, as it has been claimed that religious and cultural clashes and fault-lines are, and will be, the most probable cause of conflict after the Cold War, so attention has focused on relations between Islam and both Russia and Western powers, while, in addition, Islam is also an issue in Central and South Asia, sub-Saharan Africa and the Balkans, an apparently bland observation that is not intended to lessen its capacity for causing conflict. For example, Russian concern to restore its influence within the states that were formerly part of the Soviet Union will continue to lead to conflict between Russian forces and Islamic fundamentalists, at present in Tajikistan and Kirgizstan, and possibly in the future in Uzbekistan.

Islam is not a united force, but there are common themes of assertiveness. Much focuses on the vocal (but by no means universal) rejection of the materialist civilization and pluralist values associated with the West, as well as its political and economic views, most particularly American support for Israel. Thus, in the event of future conflict, there will be an ideological clash that will make moderation in conflict and compromise in peacemaking difficult: in South Asia, the nuclear armaments of Pakistan and India are seen as Muslim bombs and Hindu bombs respectively – an approach that does not encourage disarmament.

The lesson of the Arab–Israeli conflict is very important here. Although the well-trained, ably commanded and Western-equipped and organized Israeli military was able to emerge victorious from individual wars, especially the Six-Day War of 1967, it found it impossible to secure a triumphant long-term outcome to the conflict, and this suggests a parallel in the event of any future American–Islamic clash. Unlike Israel, however, the USA does not have any contiguous boundaries with Islamic powers, and does not contain a substantial potentially hostile Islamic population (as Israel does within its pre-1967 boundaries, let alone outside them). This may encourage a sense of the controllability of war on the part of American policy-makers: the 'we-can-fire-cruise-missiles-and-switch-channels' syndrome. However, as the Iraq conflict of 2003 showed, full-scale war poses problems for American political and military capability, not least that the American expeditionary force posture is not designed for long-term occupation tasks and commitments.

The contrast in time-frames, and political cultures, between the USA and non-Western powers will not be the same for wars between non-Western states that do not directly involve Western powers, for both military and political reasons. First, the tempo of military activity in most wars between non-Western states will probably be lower, and thus the rate of depletion of military assets will be slower. Secondly, political expectations and needs for a rapid result, while still pressing, will be less powerful. In the next chapter, we shall turn to how wars will be fought. Suffice it to note here that the majority of international wars in the future will probably continue not to involve the leading military powers directly, although their weapons, expertise and interests may all play a role, and are especially likely to do so in Asia.

Africa

These wars are likely to be particularly common in sub-Saharan Africa, the continent of war, a region where instabilities that arise in part from the legacies of colonial rule, especially the mismatch of political and ethnic boundaries, will continue to be exacerbated by a resort to force that, in part, reflects the inability of states to develop democratic mechanisms for the internal conduct of politics, as well as by the difficulty of moving past ethnic constructions of politics, the pressures of competition for resources and the availability of large numbers of soldiers of fortune. Post-colonial political structures and practices have failed to ensure peace, prosperity and adequate or fair government. Indeed, most African states preside over public systems, whether of education or transport, health or finance, that are weak when they are not in crisis. Their governments generally devote little attention to these issues; instead they put a greater emphasis on the politics of power, both domestic and international, being unwilling to yield power peacefully, and also being keen to pursue foreign interests by war, both, for example, being characteristic of Robert Mugabe's policies in Zimbabwe.

While such preferences continue to be widespread, it is difficult to anticipate much long-term peace in sub-Saharan Africa. The colonial powers have gone, but instead countries are 'colonized' by regimes and their allies, for example the late President Mobutu in Zaïre; or Mugabe, who responded to democratic opposition by encouraging his security forces to act violently, in June 2003 meeting street demonstrations with a show of military force as well as the liberal use of beating and tear gas; while, at the same time, the economy collapsed. Mugabe was far from alone in his use of violence to maintain domestic control; although it is not clear, in the case of some regimes, how far repression was ordered by 'the government' and how far it was the product of elements of the regime operating in accordance with their own agenda. This was true, for example, of the pro-government gangs that used violence against opposition figures and demonstrations in Haiti in 2001–02, as well as of the comparable groups in Myanmar in 2003; although in the latter case central direction seems probable.

African regimes are frequently identified with tribal groups, although they tend to take their own, non-consultative, definition of tribal interests. States that comprise a number of often clashing ethnic groups, such as Angola, Congo, Ivory Coast, Kenya, Rwanda, Burundi and Nigeria, will continue to be particularly unstable, and this will also affect their relations with each other. This instability may spread to South Africa, where tribal identity remains strong and the role of the state as regulator and employer will encourage efforts to control its power. Instability in sub-Saharan Africa also affects relations with neighbouring countries. In addition, attempts by states such as Nigeria and South Africa to act as regional hegemons will cause instability in Africa.

Latin America

If, as already suggested, there will also be international instability leading to warfare in parts of Asia, the situation is less clear in Latin America, where American dominance may continue to restrict wars and to restrain the activities of the military and of military governments. In 2000 American and Brazilian pressure on Paraguayan military leaders led them to thwart an attempted coup. The fact that since the 1940s wars have not been normative in Latin American politics may be very important for the future; serious disputes, such as frontier disagreements between Argentina and Chile, have been settled without recourse to war, and there seems little reason to imagine that the situation will change.

However, alongside this optimistic vista, there is another that suggests that the pattern of intervention in the instabilities of neighbours seen in Central America in the 1980s, with Nicaragua, El Salvador, Honduras and Guatemala, may be repeated and may become more serious. The essential acceptance of state sovereignty that has characterized South American power politics might be eroded by a convergence of rising domestic pressures, not least drug-financed insurrection and terrorism, with opportunistic responses to the instability of neighbours. However, there are fewer contested frontiers than in

Africa, while the Latin American militaries are concerned with stability within their own countries, rather than war with neighbours.

It is unclear whether conflict in Latin America, be it international or domestic, would have a major impact on world politics, although such a statement begs the issue of how such a concept is to be defined and removed from subjective criteria. Nevertheless, in terms of geopolitics, the absence of close links between the Latin American states and rivalries between major global powers helps lessen the importance of conflict in Latin America. There is nothing comparable to the Egypt– or Syria–Soviet Union and Israel–USA relationships that helped to keep Middle Eastern conflicts at the forefront of attention during the Cold War, nor to American sponsorship of Pakistan and Soviet links with India in the same period. In South America, communist attempts to build up powerful insurgency movements failed. American dominance remained powerful, whether the tide moved towards dictatorial or democratic regimes, and within most states there was no key resource such as oil to provide occasion or funds for conflict, although in some states drugs now provide that resource.

Looking ahead, it is difficult to anticipate any changes: Latin America may well be a vulnerable 'flank' for the USA, but the challenge may rather arise from Latin America's relationship with sources of tension within the USA, such as economic competition, immigration and drug consumption, than from military or political factors. The extent to which Cuba's strategic threat under Castro does not undermine American security at present suggests that it is inappropriate to envisage serious problems if, for example, Bolivia was ruled by an actively anti-American government. However, concern about failed states and their possible use by terrorists will accentuate America's support for stability in Colombia and may also lead to greater intervention than hitherto in attempts to overthrow Chavez's pro-Cuban populism in Venezuela.

Europe

Discussion of the USA, and due acknowledgement of the role and importance of its military power and political leadership, should not lead to a neglect of the possible future importance in conflict and confrontation of both allies and alliance structures. Here the Americans face a very uncertain future because the end of the Cold War, while leaving the USA's hegemony even clearer and enabling it to pursue new alliances, particularly in Eastern Europe and Central Asia, also weakened the cohesion of its existing alliances (though there had also been serious tensions in earlier periods: for example, the lack of European support for the Vietnam War). The lack of cohesion in existing alliances is more true of Europe than East Asia, because fear of China and North Korea plays (and will continue to play) a major role there, especially for Japan, South Korea and Taiwan, encouraging them to rely on the USA for protection. And in so far as Chinese–American tensions create a new bipolar geostrategic confrontation, it will focus on East Asia.

In Europe, there is a continuing threat of instability in the Balkans and of confrontation with hostile powers in North Africa, as well as a desire for stability in the Middle East, but this will not cause a maintenance of cohesion in relations with the USA comparable to the Cold War, because the logic of NATO is challenged by the pretensions of the European superstate, and, specifically, because of a deep anti-Americanism in France, the state whose political conceptions will continue to be most influential in the EU. Thanks in part to its close links with Germany, France is far more influential in the EU than Britain, the major state that is happiest to adopt, and adapt to, American views, and American hopes that closer British engagement with the EU will lead to an EU that is more responsive to the USA are naïve, although they largely account for American pressure for such engagement.

Furthermore, the expansion of NATO into Eastern Europe threatens to weaken its cohesion, certainly in the event of any crisis involving commitments. For example, it is unclear whether guarantees to former communist states would be supportable in the event of their being challenged by Russian-backed opposition movements that employed violence but were not supported by Russian troops. In the future this may be a particular problem in the Baltic states, especially Latvia, and to a lesser extent Estonia, as both have large Russian minorities. Similarly, NATO interest in the Caucasus may cause difficulties, with the ambitious views of Turkey, a NATO state and would-be EU member, not shared by other countries in either alliance.

The EU itself showed in both the two Gulf crises and Kosovo war that it is unable to fulfil the expectations for political, let alone military, action that its proponents claim. This is likely to remain the case, and this weakness will affect the military situation, both in the eastern Mediterranean and more generally; the consequences including strains in American multilateralism, not least as the EU's pretensions to independent policy-making clash with the

USA. The projected 60,000 strong EU deployment force is intended in part to meet American demands that European powers take on a heavier defence burden, but such a force is also intended to provide an out-of-NATO strike capacity that may lead to policy disagreements with the USA. The Americans want the EU to be a reliable ally on American terms – in short complementary to the European arm of NATO – but that is wishful thinking, and this tension became more apparent in 2002–03.

It is possible to advance a more optimistic scenario, to emphasize the continued relevance of NATO and transatlantic links, and to argue that arms controls and improved relations with Russia, and the expansion of NATO and the EU, will lead to a more secure and stable Europe in which peacefulness will be normative. The latter scenario, however, depends on overcoming traditional practices and concepts of state interest, and also on ensuring that when events help create crises, for example between Greece and Turkey, that it is possible to maintain a systemic coherence, in short to prevent a breakdown of relations within Europe.

A decade ago, the collapse of the Soviet Union and its alliance system interacted in Europe (and elsewhere) with a Western globalism that seemed to leave scant role for traditional spheres of interest, and was thus a challenge to established notions of stability. However, the problems of such a policy have encouraged more caution of late, as shown in the acceptance that Russia will maintain power by brutal means in the northern Caucasus. Furthermore, the very traditional assumption of geopolitical and geoeconomic blocks that the EU represents will probably encourage a desire for stable neighbours and neighbouring blocs, although this policy has been less than successful in the Balkans, while there has been considerable ambivalence towards Turkey. The EU depends on Russia managing a successful transition to liberal capitalism and political pluralism, and on the absence, there and in the EU, of a populist nationalism that threatens, intentionally or otherwise, to lead to conflict with neighbours. Developments in former Yugoslavia underlined the risks both of conflict within European states and that these will draw in other powers, but it is probable that Yugoslavia was unique in Europe. The peaceful division of Czechoslovakia into the Czech Republic and Slovakia in 1993 removed one threat of civil unrest, and although there are powerful ethnic and regional tensions elsewhere (for example in Romania) it is unclear that they can lead to war; more likely, if they worsen they will lead to large-scale lawlessness with any foreign intervention being under the guise of humanitarian intervention or peacekeeping.

An understanding of the dangers of a weakening of NATO through the attempt to develop an independent European security identity serves as a reminder of the future role of contingencies. If the military history of, say, the 2020s is seen as likely to be different, depending on whether NATO remains American-led and powerful, or not, with consequent influences on doctrine, strategy and force configuration, then the political contingencies that might determine this outcome become far more important. They include political pressures within both the USA and Europe, as well as the impact of crises on relations between the two, for example conflict in the Balkans, Middle East or Far East. The Iraq crisis of 2003 saw both American nativism and European hostility to America, each encouraged by xenophobic tendencies, and was a reminder both of these pressures and of their dependence on contingencies. Over the past century, wartime cooperation in the North Atlantic region was crucial both to the defeat of autocratic hegemonic tendencies in Europe and to the security of the USA. Although it is unfashionable (and perhaps somewhat controversial) to say so, Western civilization was saved three times by the USA in the twentieth century; though Germany in 1914 was not a threat comparable to those posed by Hitler and the Soviet Union, while Britain played a far greater role than the USA in World War I and in 1939–41. If a comparable challenge recurs, it should not be assumed that it will be possible to defeat the challenge by the same means, for such a scenario depends on continued effort and on circumstances, and is in no way inevitable.

Conclusion

To refer to a comparable challenge may seem alarmist, but it is worth remembering that, hitherto, all postwar periods have been interwar. From the perspective of critics of the USA, its power is in part an aspect of the degree to which the menace of an overly strong and vigorous hegemonic power has a habit of responding to domestic strains elsewhere and to international vacuums. Supporters of the USA, and those who regard its power as fundamentally benign, will see both as a potential challenge to its position. If such strains and vacuums are seen as inherent in political systems, then it becomes less easy to be confident about the future. Furthermore, hitherto all international systems have had a dynamic quality that has eventually entailed paradigm (fundamental) shifts. It is probable that the latter will occur again; but it is unlikely that such a shift could be accomplished without major disruption, possibly including conflict. Irrespective of this point, the notion that future war will somehow be the fate of 'rogue' states and barbarous areas that cannot sustain statehood, while the West can choose isolation or low-risk 'peace-keeping', is overly complacent. An understanding of the multiplicity of reasons why war may break out is important, as an understanding of the causes and, still more, purposes of conflict helps explain its conduct. It is to that which we will now turn.

4 Future Conflict

> All the high-tech weapons in the world won't transform the US armed forces unless we also transform the way we think, train, exercise, and fight.
>
> (Donald Rumsfeld, 2002)

An understanding of the diversity of future wars helps explain the range of future conflict. There will be no one type of war, and thus no one way of waging or winning war. Military goals will vary greatly, as will political contexts, and the two are closely linked. However it is also important to note a degree of autonomy: while war may be the pursuit of politics, the institutions for, and demands of, conflict have needs of their own that do not readily respond to political requirements. We will begin by looking at 'high-tech' war, but then continue by stressing the need to consider other types of conflict.

High-tech warfare

Much of the assessment of future high-tech war depends on two concepts: synergy (profitable combination) and information warfare. The former suggests that future success will hinge on the ability to achieve a successful synergy of land, air and sea forces, which requires the development of new organizational structures, as well as careful training of commanders and units, and appropriate systems of command, control, communications and information appraisal and analysis. These notions reflect not simply an awareness of the benefits for interdependency of ground, air and sea brought by technological developments but also the experience of conflict from World War II on, especially the important examples of successful air–land, land–sea, air–sea, and air–land–sea cooperation. It became clear in particular that rather than being an optional extra, air power was vital to operational success on land and sea.

Furthermore, the experience of land warfare has shown that technological advances worked best only if secured in a combined arms approach. Thus, in World War II, using tanks alone to try to revolutionize warfare was found to be of limited value in the face of 'counter-tank' practices, both the use of anti-tank weapons and the employment of tanks in mobile defence; and the emphasis on the cooperation of armour with infantry and artillery that affects current thinking about future land conflict in large part stems from experience, and should be a warning against continued popular fascination with armour alone. To interject a personal note, an American tank colonel in conversation in 1997 responded to my query about the possibility of conflict over Taiwan by remarking that Taiwan was too small for his unit to operate in; but the sense that the real world should conform to weaponry is flawed.

Doctrinal developments have also reflected particular conjunctions. The USA's promulgation of the doctrine of the AirLand Battle in 1982 was designed to offset the particular problem of Soviet numerical superiority in Europe; but, looking ahead, it enshrined the notion that a successful combination of firepower and manoeuvre would be a vital addition to positional defence, and that this would require an effective synergy of land and air. In addition, AirLand looked ahead in its emphasis on a level of conflict, and thus planning, between tactics and strategy: this operational level had already played a major role in Soviet planning and is likely to play an even more important role in future Western military doctrine and planning, not least because it reflects the modern transformation of strategy.

Advanced technology, and a sense that technology would continue to advance, and that this has to be planned for, greatly contributes to ideas of synergetical warfare, which reflect an awareness of the need for a more sophisticated command and planning environment, and also a need to do more than simply respond to the possibilities created by new weapons. Instead of treating these in isolation, their impact is to be multiplied by careful cooperation.

There is no shortage of recent advances in weaponry, nor of plans for further new weapons. The Americans now use stealthy attack aircraft able to penetrate opposing, integrated air defences, thus moving the dialectic of offence and defence; as well as employing 'smart' guided weapons fired from stand-off platforms. Laser- or global-positioning system-guided projectiles and programmed cruise missiles are employed to attack ground targets, while advanced aircraft are able to win air superiority. Looking ahead, there has been discussion of the use of high-speed aircraft capable of 'skipping' on the upper atmosphere and of transporting troops anywhere in the world within two hours.

Such technology has to be impacted within new processes to ensure effectiveness, not least at the expense of the possible use of advanced weaponry or 'anti-weaponry' by opponents. Coordination is now made possible by computer networking, information assessment through spy satellites and other important sensors, such as AWACS aircraft and the use of global positioning systems. At the same time, the importance of these information supplies ensures that effort has to be put into making them secure from both interception and disruption, and the risk of the latter increases as real-time information is integrated into decision-making at lower unit levels, for example targeting by individual tanks, guns and soldiers.

This serves as a reminder of the potential vulnerability that comes from improved systems; vulnerability that needs to be built into any assessment of capability. However advanced, modern communication systems are vulnerable to a number of disruption mechanisms, as well as to surveillance, as the British discovered in Kosovo in 1999, and this is especially true in the confusion of combat. In addition, information coordination is more difficult in practice than in theory, especially in high-tempo situations. This affects not only the movement of information to individual units, and its analysis with reference to their needs, but also the relationship between planning and operations. As the speed of communications will increase, and this will also be true of opposing forces, so the pressures of simultaneity will be exacerbated. It is by no means clear how far command and control, which ultimately depends on human decision-making, will be enhanced or compromised by the massive and rapid flow of information; while the visual understanding of quantities of information, much of it in three dimensions, will also pose problems.

Furthermore, it is unclear whether the resulting control will focus on commanders able to direct the moves and fire of distant vehicles, platforms and soldiers, or on the latter, able to have continuous and instantaneous access to all relevant combat-zone information. This is an aspect of a more general transition from the massing of forces and firepower that characterized conventional warfare, through the massing of firepower using dispersed forces, such as rocketry, to the ability to use precision-guided munitions in order to achieve mass effects without having to mass forces or firepower.

The maritime dimension

In Western planning, there will also be an emphasis on the sea as a sphere for manoeuvre and as a base area. This will reflect the problems that civilian opposition and insurrectionary movements pose for the use of land as a military base, and the geopolitical shifts produced by the end of colonial bases, and by political sensitivities over the deployment of troops in the territory of allies: in the Kosovo crisis of 1999, there was considerable hostility in Greece, a NATO power, to the use of the country for the transhipment of NATO forces and supplies. As a consequence of these changes, the geopolitics of alliances have changed: a shift that can be seen as a counterpart to the RAM already discussed.

There is no reason to believe that this process will be reversed, but rather every reason to anticipate that it will be accentuated. It may, in time, affect the use of Western European facilities by the USA, and American concern about this led, in 2003, to the development of bases in Eastern Europe, where states such as Bulgaria were seen as more pro-Western and ductile than Germany, as well as closer to the sphere of American concern in South-West

Asia. The expansion of the EU to include states with neutralist traditions, such as Austria, Finland and Sweden, may be followed by further expansion to include more, for example Malta or (Greek) Cyprus, which will encourage a hostility to the use of the EU as a base area for American forces or operations.

The possibilities of sea-based forces have been enhanced by new technology. Basing aircraft on ships and, later, ballistic missiles on submarines was followed by the development of sea-based guided tactical missiles; although it is difficult to employ them to provide a volume of fire sufficient to provide continual support for units engaged onshore. This problem reflects limitations in the firepower and ammunition storage of modern warships. Active naval construction programmes continue, and not only in the USA. Launched in 1998, France's 40,000-ton nuclear-powered aircraft carrier *Charles de Gaulle* is the largest European warship in decades. It has faced serious operational problems, particularly related to its rudder, while the availability of only one carrier raises the question of how best to maintain capability when it is not on station, especially if under repair. In 1998 Britain announced plans to build two carriers of a similar size, and Italy approved a small carrier with amphibious assault capabilities.

Aside from the sea as a platform for mounting bomb and missile attacks, there will continue to be an emphasis on the potential for amphibious operations, whether in attack, in reinforcement or for the defence or withdrawal of interests and people. Such operations offer the seizure of the initiative, manoeuvrability and combination of different arms sought by modern strategists. Their value will remain high because so much of the world's population and economic power is located on, or close to, littorals (coastal areas), and because so many states have coastlines, and are therefore vulnerable. Thus the sea was used as the initial approach route for the British intervention in Sierra Leone in 2000; while the American Marine Corps developed a doctrine focused on what was termed Expeditionary Maneuver Warfare. At the present time, there is also much work on the development of medium-sized amphibious vessels. A stress on amphibious capability will also be important because only a small number of states possess and will possess this. Indeed, a sea-based strategy will be of use to naval powers not only because they will have the capability but also because it will enable them to emphasize their distinctiveness, and thus to gain an important psychological advantage.

The relative safety of the sea will maintain the contrast between military capability on land and at sea that will be so important in future military capability and conflict. Irregular forces operate far less at sea, and this will continue to be the case. There will continue to be high levels of piracy off parts of South-East Asia, especially Indonesia, and, to a lesser extent, in the Caribbean, and these may spread, for example in African coastal waters; but such vessels will not be able to threaten the warships of major powers, and nor indeed will they seek to do so, because the economic rationale of piracy will depend on avoiding such confrontation.

Furthermore, although modern warships and those on the drawing-board are soft-skinned compared to their predecessors, and certainly not armoured behemoths, which helped account for the serious damage suffered by the USS *Cole* in Aden from a suicide boat, their protection will be enhanced by sophisticated surveillance systems linked, in particular, to anti-aircraft and anti-missile weaponry, and their range will be extended by aircraft, helicopters and missiles. At sea, it is, and will continue to be, easier to distinguish and assess other units than on land, and, as a consequence, it will be possible to avoid the situation on land in which guerrillas will be indistinguishable from the civilian population.

The sea, in short, will be known, although the application of stealth technology to warships will pose a problem. As a consequence of the range provided by aircraft, helicopters and missiles, ships will be far better able to mount attacks on hostile powers than their predecessors, while the 'transition costs' of amphibious operations will be greatly lessened as the point of contact (and emphasis) of sea-based attacks will cease to be simply the coastal landing zone: the littoral will replace the coastline. Thus, the US Marines have switched their focus to ships-to-objective manoeuvre, a system that was intended to overcome the traditional emphasis on seizing and exploiting bridgeheads. Instead, by going directly to the target, this new approach is seen as a way to bypass defences and to maintain the initiative.

Naval power will offer reach, mobility and logistical independence, providing a dynamic quality that will be lacking from fixed overseas garrisons, and will thus offer a military substitute for colonial possessions. This will be particularly appropriate for the world's leading naval power, the USA, because its political culture does not make colonial rule acceptable. This is also true of the other major naval powers (Britain, France, Japan, Russia), with the exception of China: the use of naval power in support of a seizure of Taiwan would be presented by the Chinese as a reunion, but it would also have, at least in part, an imperialist character. Elsewhere, however, it is on land that China pursues colonialism – in Tibet and Xianjiang – not at sea.

The expense of major vessels and of naval infrastructure is such that naval power will become even more of a high-cost exercise, that, in part, reflects the need for very advanced electronics, which are largely a response to the extent to which surface vessels became more vulnerable last century as a consequence of the development of submarines and of air power. It has become necessary to devise countermeasures, and also to ensure that the various forces that can operate on, over and under the sea combine effectively; this will become even more the case if the hopes of protagonists of sea-to-land capability are to be realized.

The limited number of states with any major naval effectiveness suggests that the sea will essentially be used for sea-to-land operations, rather than contested between marine powers. Nevertheless, the possibilities of naval conflict have been enhanced by the increase in the number of naval powers, as well as by the naval plans of a number of states. Naval shipbuilding has developed in newly industrialized countries, especially Argentina, Brazil and India, there has been an arms race in East and South-East Asia, and wealthy states that lack large navies have proved willing to buy advanced naval vessels, so that, for example, Iran has acquired sophisticated Russian submarines.

It is possible to point to particular issues that may lead to the use of such vessels. For example, disputes over islands and territorial waters in the South China Sea involve the states that border the sea, especially the struggle over the Spratlys. Episodic fighting in the sea since

1974 involving China and Vietnam, and confrontations including, in addition, Malaysia, the Philippines and Taiwan, have fuelled naval build-ups in the region, although hitherto they have led to clashes and not to war.

It is unclear how far this is likely to remain the case. The Western Pacific/East Asian littoral zone is one of the most volatile regions in world power politics, in geopolitical terms a fault-line, akin to that between two tectonic plates, in this case the Pacific, which is currently under American control, and the East Asian littoral, where American interests are challenged. Such an instability can be seen as likely to restrain war, as the close proximity of heavily armed forces might lead to a caution bred of fear about likely escalation, but it could also lead to war by accident, or encourage the search for apparently 'safer' topics for conflict. Naval clashes in the South China Sea might well serve as a prime example of these, enabling China to win a contained conflict with a state far less powerful than the USA. There are also in this area the resource interests mentioned in the previous chapter, in particular oil.

The increase in the number of states with submarines may ensure that submarine warfare and anti-submarine capability will play a greater role in future confrontations than in the conflicts of the 1990s. The security of deployment enjoyed by Western forces in the Kosovo and two Gulf Wars (as earlier in the Korean and Vietnam Wars) may not be readily replicated, and a modest outlay in submarines by hostile powers would oblige Western forces to regard their deployment routes as an active combat zone. This would have an important impact on naval resources, now severely stretched due to the rundown in the number of units. It is worth thinking back to the Falklands War of 1982 and considering how many warships the British would have had available for operations round the Falkland Islands had they also had to protect routes right the way back to the British Isles from the threat of Argentinian submarine attack.

Such a threat may well serve further to encourage long-distance deployment by air – a capability identified as seriously deficient in European forces in the early 2000s – but this deployment requires safe landing-zones; while, on the ground, aircraft are very vulnerable to attack, and in addition the use of aircraft is restricted by the weather. Reliance on air also makes it even more likely that future distant deployments will not include sizeable armoured forces, as tanks require much space and a formidable support and supply system, including mobile repair facilities. Tanks can be moved in aircraft, but it is likely that an enhanced reliance on long-distance deployment and resupply by air will contribute to the growing marginality of armour.

This will be a marked contrast with the situation during the Cold War and in the two Gulf Wars. The contrast will also mark a major shift in popular assumptions about the character and weaponry of land conflict. However, the importance of US armour in the second Gulf War, not least in the face of the Iraqi use of rocket-propelled grenades against US vehicles, may lead to a rethinking of this point. It certainly casts doubt on the idea of reliance on the Stryker wheeled combat vehicle, as, while capable of firing missiles, and thus providing precise striking power, as well as mobility, such vehicles lack the combat-durability of tanks. The second Gulf War certainly throws open the question of the best force structure for expeditionary forces.

The Gulf Wars

If the sea is likely to be of primary importance in terms of a base for force projection onto land, and for logistical and transport capability, then it is still necessary to consider how far future land conflict will reflect present patterns. Much of the problem arises from different judgements of the latter, and, as planning and training for future wars in large part reflect the 'lessons' of the past, then the controverted nature of the latter is particularly important. This is particularly true of the Gulf Wars of 1991 and 2003, the Kosovo War of 1999, and the Afghanistan conflict of 2001.

Learning from past performance is important as it offers an assessment of current capability, and, in the absence of evidence of such performance, it is difficult to judge capability. For this reason, the continuing ability of states to understand their relative position may depend on the frequency of conflict in the future, which will cause a significant problem because large-scale conflict involving major states is (hopefully) likely to be infrequent, and thus it will be very difficult for militaries to realize deficiencies in time to do anything about them: major war may well be an event, rather than, as in 1689–1713, 1739–48, 1792–1815 and 1914–45, a sequence. This situation will be exacerbated by the brevity of large-scale conflict involving such states.

The Gulf Wars of 1991 and 2003 have played a particularly important role in current discussion because they led to what could be presented as clearcut victories and were played out in the full blaze of publicity; and also because offensive campaigns have always attracted more analysis than their defensive counterparts. Furthermore, the Gulf

Wars appear to be a model of an important type of future conflict because they opposed the USA to a 'rogue' state. These wars were widely presented as a triumph for technology, and for doctrines based on advanced weaponry. As pictures of precision bombs and missiles destroying their targets were endlessly repeated on television, the wars helped to resurrect popular and air force confidence in the capability and impact of air power following its apparent failure in the Vietnam War. The Gulf Wars also tested the doctrine and practice of the AirLand Battle, and did so at the moment when American analysis and preparation for what was termed the RMA were coming together.

In the Gulf Wars, the Americans employed high technology in reconnaissance, offence, defence and communications, and all were hailed as anticipations of the future nature of conflict, not least as advertisements for particular weapons systems that would be employed to fight future wars. Reconnaissance was not separate from offence or defence, but was closely linked to the weapons available for both, so that satellite surveillance was employed to track Iraqi missile launches, as well as to guide individual American units. Near-real-time information and communication provided many more opportunities for individual units on the ground to take decisions, and led to pressure for real-time information. American tanks successfully employed precise positioning devices interacting with American satellites in using a global positioning system, while satellite imagery was responsible for the rapid production of photo-maps. Information was also important to the missile assaults by Tomahawk cruise missiles, which made use of the precise prior mapping of target and traverse in order to follow predetermined courses to targets that were actualized for the weapons as grid references. Thus, digital terrain models of the intended flight path facilitated precise long-distance firepower.

Firepower was also enhanced by other methods. Thermal-imaging laser-designation systems were employed to guide bombs to their targets, and tanks also made effective use of thermal imaging sights. The use of such new technology permitted a reconfiguration of older weapons-platforms. For example, in Afghanistan in 2001, American B-52 aircraft, originally designed for international strategic strikes, were successfully armed with joint direct attack munitions so that they could provide effective close-air support. Other high-tech weaponry included B-2 Stealth bombers, which were able to bomb Baghdad, one of the most heavily defended cities in the world, with total impunity and considerable precision. Stealth technology did indeed minimize radar detection, while in-flight targeting greatly enhanced the flexibility of air power, and helped it operate as a readily responsive part of the multiple-source information-based warfare the Americans employed. The range of weapons included Patriot anti-missile missiles, used to protect bases from Iraqi attack, unmanned air vehicles, electronic counter-measures and fuel-air explosions.

The attack weaponry was supported by sophisticated and complex control systems designed to help maintain the pace of the attack. Compared to earlier conflicts, target acquisition and accuracy were effective, although the defensive character of the target, especially the entrenched Iraqi forces in 1991, and the desert nature of the terrain greatly assisted. The Iraqis were defeated twice with heavy casualties and, proportionally, even greater losses of equipment, while their opponents lost very few men or equipment.

Thus the Gulf Wars appeared to show the way ahead. 'Rogue states', however well armed, would be defeated by well-trained American or American-led forces capable of using advanced weaponry in an effective fashion, and to a timetable. Yet, the wars were capable of a very different analysis which also looked ahead to suggest that the nature of future warfare would be less comforting for protagonists of the RMA and for those who suggested that American hegemony would be relatively unproblematic. First and foremost, there were doubts about the effectiveness of high-tech weaponry, and thus about the impact of a future capability advantage based on technological strength and operating through such weaponry. For example, subsequent analysis of the 1991 war indicated that the Stealth planes, Tomahawk and Patriot missiles and laser-guided bombs did less well than was claimed at the time. In particular, their much-lauded accuracy was less manifest in combat conditions than had been anticipated, especially, in 1991, that of the Patriots, as also of the Soviet-supplied Iraqi Scud missiles.

Furthermore, command and control proved unequal to the fast tempo of the conflict, unsurprisingly so. Indeed, the system buckled under the combined strain of its very complexity and of this tempo. AirLand battle proved more difficult in practice than in theory, not least due to the problems of synchronizing air and land forces under fast-moving combat conditions, and this led to 'friendly fire' fatalities. The deficiencies of air and missile attack attracted most attention, but other limitations were also revealed. For example, in 1991 on the ground, supporting artillery fire suffered from deficiencies, including a lack of adequate ammunition and poor integration at unit level.

In addition, analysis of the 1991 war suggested that factors other than weapons and their use were crucial. The 'Instant Thunder' plan devised by John Warden to employ a strong air attack on Iraqi centres of gravity in order to paralyse Iraqi command and control systems and lead the Iraqis to withdraw without the need for a ground offensive proved over-optimistic when tried in 1991. Much of the success of the Allied coalition then (and in 2003) was arguably due not to respective weaponry but to Allied, principally American, fighting quality, unit cohesion, leadership and planning, and to Iraqi limitations in all four, and in other respects. In addition, in 1991 the Iraqis surrendered mobility and the initiative by entrenching themselves to protect their conquest of Kuwait; while in 2003 they rested on the defensive, apparently confident that they could draw the USA into urban warfare where American technological advantages would be of limited value. In both wars, the Allied task and the Iraqi response were suited to what has been seen as the American systematic production-line approach to warfare, and attention was not drawn to the limitations of the latter, in large part because Iraqi deficiencies were ably exploited by the Americans, and also because the Americans won. Although the margin between success and failure in war is often narrower than is appreciated, and it is all too easy to explain as inevitable what, in fact, was far from being so, this was not the case in the two Gulf Wars. In military terms, both offensives were dramatic successes; while many of the problems that had been feared in the battle-zone, such as chemical attacks, or elsewhere (e.g. Iraqi initiatives to bring Israel into the conflict, or serious terrorist attacks) did not occur.

In the 1991 Gulf War, there were claims that a Western failure of nerve allowed Saddam Hussein to survive the crisis; in short an echo of the argument that domestic, political and media pressures subvert the Western military: a claim much voiced in analysis of the Vietnam War. In practice, there was a failure to match military plans to the desired outcome of the overthrow of Saddam, as well as the problems of coalition warfare, and a concern that if Iraq imploded, Iran would become too powerful. However, no comparable lack of resolve was shown by the Israelis in their operations in Lebanon over the last quarter-century; instead, major operations in 1978, 1982, 1994 and 1996 were thwarted not by Israeli politics, but by the intractable nature of the opposition. At one level, a high-tech 'solution' was possible – Lebanon could have been turned into a nuclear wasteland – but such a war-induced holocaust was not an option, and instead the Israelis learned the deficiencies of waging war using advanced conventional weaponry. In 2003 Saddam was overthrown, but it proved very difficult to translate battlefield success into uncontested political control.

The Kosovo conflict

Problems with advanced weaponry were also extensively discussed during, and after, the Kosovo War of 1999: a conflict that was fought out in the full glare of attention, and amidst considerable controversy about the effectiveness and potential of particular weapon systems. Reiterated NATO claims about the destructiveness of air power proved greatly misleading. Far more Serb tanks survived than had been anticipated, and this suggested that the use of air power against less prominent units, for example mortars, machine guns or individual soldiers, would have been even less successful. The Serbs, employing simple and inexpensive camouflage techniques, succeeded in preserving most of their equipment, and their eventual retreat seems to have been due to Russian pressure, to the maintenance of NATO cohesion and to the threat of a NATO land attack, rather than to the lengthy air offensive which involved 10,000 strike sorties. However, the impact of the latter on the economic interests of leading members of the Serbian elite was serious, while the ability of the NATO air assault to strike with near impunity greatly increased its psychological impact.

The report produced by the British National Audit Office in 2000 on British operations in Kosovo the previous year depicted a series of serious limitations, including major shortages of aviation tradesmen and specialists for the RAF, as well as major operational problems: on cloudy days, the planes could not identity targets and were grounded, although, ironically, this prevented an excessive depletion of guided bombs. In addition, a considerable number of bombs mounted on aircraft were unable to survive the shock of take-offs, while many missiles carried by Royal Navy Harrier jump jets became useless after a few sorties due to heat and vibration damage. There were also serious problems with British communications: enemy forces were readily able to monitor British radio communications, and the army also lacked secure voice and data communication links to the UK. A lack of British lift capacity led to a reliance on Russian-built Antonovs hired from private contractors, the use of which was dependent on Russian certification. The SA80 rifle, the main British infantry weapon, was found to be faulty. In addition, there was a shortage of medical supplies and of accommodation. Furthermore, British Tornedo GR4 jets were reportedly unable to drop precision-guided bombs effectively. The jets also proved deficient in British operations over Iraq due to the consequences of heat, thus offering an ironic warning about the deficiencies of globalism, for state-of-the-art weaponry was assumed to be effective all over the world.

At the same time, the series of conflicts in which the USA (and Britain) engaged from 1990 permitted a rapid evaluation of weapons, commanders and units in combat conditions, and, although the process of learning always faces the problem that inappropriate lessons may be learned, it did help shape the application of new technology. Thus, the very serious impact of bad weather in operations against the Serbs in 1999, helped lead to a focus on the issue, which had a number of consequences, including an understanding of the value of global positioning system-guided weapons, which offered all-weather attack, as opposed to laser-guided ones.

The air assault in the Kosovo conflict offered scant guidance to the likely nature of fighting on the ground there had the Western forces attacked, and should not be used to proclaim the obsolescence of ground forces seeking to maintain, or contest, territorial control. It is of course possible to speculate as to the likely effectiveness of an Allied land attack on the Serbs in Kosovo in 1999, had one been mounted. It would certainly have posed a question-mark against the tactical effectiveness of Western forces; for example, close air support would have been difficult in cloudy weather, and much of the terrain and

ground cover were not suitable for Allied attack, and
instead would have favoured the defence. In practice,
there would probably have been more close-quarter fighting than in the Gulf Wars, a greater reliance on infantry
and, it seems likely, a larger use of artillery, rather than
tanks or air power, in order to weaken hostile positions.
The result, a close interaction of infantry and artillery,
would not have been anachronistic, but would have been
a reminder of the need to be cautious before assuming that
any future operations by leading military states would
necessarily focus on air power and, if on the ground,
highly mobile units. Had there been a ground attack in
1999, supply routes into and through Kosovo would have
been crucial to any Western advance, and they would have
had to have been protected. Deep-space attacking operations could not have done that, and it is important not to
exaggerate the consequences of disorientating opponents
by such moves.

The revolution in military affairs

In many respects, the infantry–artillery combination will
probably remain the crucial military option in land operations: a point that needs underlining as it is so much at
variance with the bulk of modern assumptions. The Americans showed with their success in Operation Anaconda
in Afghanistan, the artillery support could be provided by
aircraft and helicopters that were able to offer support to
troops in close combat on difficult terrain, but it is necessary anyway to stress the atypicality of American war-
making, while bearing in mind that no war is typical. In
Kosovo, there was no ground fighting between Western

and Serb forces, although there was much fighting between the Kosovo Liberation Army and the Serbs, in which the classic factors of territorial control and denial were contested on land, and the conflict did not measure up to the paradigms of information warfare, being instead closer to the notions of future warfare as a savage dialectic of guerrilla operations and brutal counterinsurgency. Prior to the Western air assault, the tasking that faced the Serb military and paramilitary forces was defined in these terms, and these have to be used in assessing their capability.

An interesting parallel was provided in 2000 when Ethiopia invaded Eritrea. As with many Third World conflicts, it is difficult to be precise about events, but the Ethiopians benefited from superior air power, better armour (Russian T-72 tanks) and greater numbers, only to find that the Eritreans fought well, taking advantage of the terrain – as presumably the Serbs would have done.

To return to the importance of infantry in the two Gulf Wars, in which attacking infantry played only a limited role, the Allies' armour was able to exploit the static or less-mobile nature of the Iraqi forces and their exposed flanks and also to use air power. It did both effectively. There was relatively little role for the Allied infantry, but there would have been far more had it not been possible to outflank the Iraqis, or, in 2003, had there been more conflict in the cities.

The approach taken in this chapter does not sit easily with much contemporary discussion, because my analysis focuses, in part, on treating the belief in the RMA as symptomatic of a set of cultural and political assumptions that tell us more about modern Western society than they do about any objective assessment of military options. This subjectivity is unsurprising because we should not assume an objectivity in military discussion and analysis somehow removed from the rest of society.

Instead, the RMA acts as a nexus for a range of developments and beliefs, including an unwillingness to accept conscription, a very low threshold for casualties, an assertion of Western superiority and an ideology of mechanization. The last is crucial for, in a machine age, worth is defined in terms of machines, and they, rather than ideas or beliefs, are used to assert superiority over other humans, as well as over the environment. Furthermore, change is the characteristic of what can be termed machinism: machines are designed to serve a purpose, can be improved, have a limited life (in the sense of being at the cutting edge of applicability) and are intended for replacement.

A stress on machines leads to an underrating of infantry, and because the stress is on moving machines this leads to an underrating of the role of artillery. Such a process is not new, and can be seen also in the modern remembrance of military success, as with World War I, where it is frequently claimed that tanks broke the impasse on the Western Front in 1918, which is an exaggeration of their role, for effective infantry–artillery coordination was far more important to the Allied success. This, however, lacked, and lacks, the glamour of the concept of the tank advance, and also does not appear as the turning-point beloved of military commentators. Looking to the past for examples, in this fashion, is valuable, as modern analysts discussing contemporary and future conflict often show the same characteristics as their predecessors.

Translated into warfare, machinism assumes that capability will vary greatly between powers, producing a ready hierarchy, that capability will change, and that this change will be easily assessed. This approach readily lends itself to the notion of perfectability, and to the concept of paradigm leaps forward. Thus the RMA is an expression of the modern secular technological belief-system that is prevalent in the West, easily meshes with theories of modernization that rest on the adoption of technological systems and serves powerful psychological needs: it is particularly crucial to Americans that this is an *American* military revolution, not least because it permits a ranking in which America is foremost, and all other powers, whether opponents, neutrals or even allies, are weak and deficient. The RMA meets the American need to believe in the possibility of high-intensity conflict and of total victory, with opponents shocked and awed into accepting defeat; rather than the ambiguous and qualified nature of modern victory. In addition, the certainty of the RMA appears to offer a defence against the threats posed by the spread of earlier technologies, such as long-range missiles and atomic warheads, of new ones, such as bacteriological warfare, and of whatever may follow. It keeps the Americans ahead. Standing for president, George W. Bush, in September 1999 told an audience at the Citadel (a military academy) that 'the best way to keep the peace is to redefine war on our terms'. Once elected, he declared at the Citadel in 2001: 'The first priority is to speed the transformation of our military'. Furthermore, the RMA can serve to support a range of Western political strategies, more particularly the doctrine, politics and military strategies, both of isolationists and of believers in collective security; although the RMA particularly lends itself to Western, more particularly American, unilateralism.

However, it is also necessary to be cautious in

suggesting too much coherence and consistency in the idea of an RMA. A less harsh view than that just outlined can be advanced if the RMA is presented as a doctrine designed to meet political goals, and thus to shape or encourage technological developments and operational and tactical suppositions accordingly, rather than to allow technological constraints to shape doctrine, and thus risk the danger of inhibiting policy. Allowing for this, much modern discussion does seem to suggest a technology-driven warfare, with a lot of the technology focused on overcoming the problems of command and control posed by the large number of units operating simultaneously, and on fulfilling the opportunities for command and control gained by successfully overcoming this challenge, and thus aggregating sensors, shooters and deciders to achieve a precise mass affect from dispersed units.

Advocates of the RMA have progressed to talking about 'space control' and the 'empty battlefield' of the future, where wars will be waged for 'information dominance' – in other words, control of satellites, telecommunications and computer networks. The American military refers to information grids and networks that must be safeguarded in wartime, and hostile ones that must be destroyed. Integrated communications technologies are designed to enhance offensive and defensive information warfare capability, while better communications enable both more integrated fire support and the use of surveillance to permit more accurate targeting; all to be achieved rapidly in accord with political and military needs, not least getting within opposing decision cycles.

Western, especially American, economic growth and borrowing capacity and American resource-allocation give substance to such ideas, because they make it easier to afford investment in new military systems; or, at least, the development of earlier ones. Furthermore, in planning for future weaponry and warfare, there has been an understandable tendency to develop the ideas and weapons of the 1990s, with the emphasis on precision, mobility and an avoidance of risk. For example, signature reduction is designed to reduce vulnerability to targeting. Many of these notions are also similar to those that earlier influenced planning for nuclear attack. Given the widespread devastation that the latter would have caused, such a remark might appear surprising, but the pinpointing of nuclear missile silos in order to provide accurate targeting for precise attacks by missiles launched from mobile units was very important in nuclear strategy.

Today, planning for future weaponry reveals a stress on cheaper unmanned platforms to replace reconnaissance and attack aircraft: whether termed UAVs (unmanned aerial vehicles) or RPVs (remotely piloted vehicles) these platforms are designed to take the advantage of missiles further by providing mobile platforms from which they can be fired or bombs dropped. Platforms do not require on-site crew and thus can be used without risk to the life or liberty of personnel, and, as a consequence, they can be low-flying, as the risk of losses of pilots to anti-aircraft fire has been removed. In addition, at least in theory, the logistical burden of air power is reduced. So also is the cost, as unmanned platforms are less expensive than manned counterparts, and there are big savings in pilot training. Unmanned platforms should also be more compact and 'stealthy' i.e. (less easy to detect), while the acceleration and manoeuvrability of such platforms would no longer be limited by G-forces that would render a pilot unconscious. In 1999 unarmed drones were used extensively for surveillance over Kosovo in order to send information on bomb damage and refugee columns; and in Afghanistan in 2001 and Iraq in 2003 armed drones were used as firing platforms. The 26-foot Predator with its operating radius of 500 miles, flight duration of up to 40 hours, cruising speed of 80 mph and normal operating altitude of 15,000 feet is designed to destroy air-defence batteries and command centres. It can be used in areas contaminated by chemical or germ warfare, and its software is programmed to be able to tell if the intended target has moved close to civilians and to suggest accordingly a change of plan.

The replacement of Tomahawks by 'jumbo cruise missiles' is similarly designed to enhance versatility without any surrender to vulnerability. Tanks are increasingly seen as obsolescent, in that their range is limited, their durability is affected by terrain and their expense is high, both in terms of purchase and of maintenance. As a consequence, there has been talk of cruise missiles replacing tanks, although this was challenged by the course of the 2003 Gulf War. At sea, aircraft carriers are seen as becoming less necessary as missiles replace aircraft, and as a consequence there has been talk of their replacement by 'arsenal ships'. The capability of such weapons is to be enhanced by designing them to work within systems or networks that bring together dispersed units and different types of weapons, for example from a number of environments: space, air, land and sea. America is at the forefront of such technology, although it is not alone in its development goals, and in order to recoup some of the cost is likely to try to sell advanced weapons to allies. It is difficult, however, to control this process. For example, in 2000 the Americans threatened action if Israel sold an airborne radar system to China that could be used against Taiwan.

It may be asked how far it is possible to reconcile such weaponry with some of the geostrategic speculation about future conflict. For example, discussion of an American search for Russian cooperation against China presupposes, at least in part, a more traditional emphasis on land frontiers and propinquity. Such a basis for operations would still benefit from manoeuvrable and deep-penetration forces, but there would be no need on the part of the Russians for sea-based platforms with their logistical and space constraints and their degree of vulnerability. Given that most discussion of future war between major powers focuses on American or, at least, Western forces, weaponry and doctrine, it is worth noting that it could also be between China and Russia or India. Whereas, in the 1960s this would have found the Chinese stronger on numbers than weapons, the situation is different now.

Furthermore, in contrast to the emphasis in modern discussion of a lack of Western interest in territorial expansion, it is possible to see a very different situation in this case, as there is a border that the Chinese have reason to want reversed. Russia gained the Amur region in 1858 and the Ussuri region in 1860: a frontier delimited by the Treaty of Beijing of 1860, which was wrung from the Chinese in a year of defeat and humiliation (Beijing was occupied by Anglo-French forces in 1860). While not as humiliating and deeply ingrained in Chinese consciousness as the loss of Hong Kong to the British, this is unfinished business. Regaining these territories would undermine the Russian position in Siberia and provide opportunities for enhanced Chinese influence and resource acquisition, whether in cooperation with, or in opposition to, the Russians.

Such a conflict might appear implausible at present, but it is not impossible in the future, and therefore has to be considered. Were it to occur, it is likely to be waged without the concern to minimize civilian and military losses that characterizes Western operations; although it would be unclear whether it would be possible to contain the conflict or whether nuclear weaponry would be employed. There is likely to be a similar emphasis on mobility, but also a greater reliance on the attritional characteristics of firepower, and a greater willingness to engage in frontal attacks (provided that there is a firepower advantage), rather than searching for a vulnerable flank as the Americans did in the Gulf Wars. The vulnerability of the Russian Far East to Chinese attack from Manchuria makes the situation very different to operating into Siberia. For the Chinese, an advance overland to the Sea of Okhotsk in order to cut off the region, followed by the capture of Vladivostok might appear a tempting 'small war', especially if Russian control of the Far East had already ceased to be effective and/or if the Russians were already heavily engaged in Central Asia and the Caucasus. The extent to which 'small wars' between nuclear powers have been made redundant by their weaponry has been challenged by the recent experience of sabre-rattling between India and Pakistan, and, rather than making such conflict redundant, it may keep it limited.

Mobility is the goal of the militaries of the future, as it is seen as both the best way to fight if conflict breaks out in established zones of tension and the best way to respond to other challenges. Thus, the Weizsäcker commission on the future of the German military that reported in 2000 recommended that the size of the German rapid deployment forces rise from 60,000 to 140,000 men. This trend towards mobility has been led by the USA, followed by Britain. In Western Europe, France and, to a lesser extent, Spain and Italy have followed suit. Elsewhere in the world, those states that seek a distant projection of their power are also pushing greater mobility. Forces need to be deployed rapidly, and then to be immediately effective.

The weaponry being developed by the USA and other states as the tools of future war is designed to ensure what are termed dominant manoeuvre, precision engagements, full-dimensional protection, focused logistics and information warfare. All of these are seen as the goals and methods of future military structures, and particular organizational forms and weapons are presented as intended to serve these ends, rather than simply moulding the structures or methods for a world of joint operations. Indeed, planners increasingly think in terms of the redundancy of traditional service distinctions, and there has been a proliferation of 'joint' organizations. In the USA these include the creation, in 1992, of an Expeditionary Warfare Division in the office of the Chief of Naval Operations and, more generally, the Goldwater–Nichols Department of Defense Reorganization Act of 1986, which strengthened the position of the Chairman of the Joint Chiefs of Staff and established a joint acquisition system. In Britain, the plethora of joints include the Joint Rapid Deployment Force, the permanent Joint Headquarters and the Joint Services Command and Staff College. Joint institutions provide powerful advocates for new doctrine and plans, such as the American *Joint Vision 2010* plan, published in 1996, which laid stress on how to obtain and use information superiority. This was followed by

Concept for Future Joint Operations (1997) and by *Joint Vision 2020* (2000). In 2002 Richard Myers, the Chairman of the American Joint Chiefs of Staff, noted that 'the rapidly changing international environment and the global war on terrorism require that we create joint capabilities more quickly. Seams between organizations must be eliminated and service and joint core competencies integrated more effectively'.

Joint institutions and planning have many uses. Given that particular weapons systems and strategies are advocated and discussed in terms of doctrines that reflect the composition and culture of particular military institutions, there is much to be said for pressing the case for unity and giving it institutional form and focus. Furthermore, traditional distinctions between the services have generally been more fluid and affected by circumstances than they may appear and are often presented, and there is no reason why there should not be even more fluidity in the future, although that is not the same as unitary tasking. The abolition of the notion of separate services may have some value for conflict between conventional forces, but there is a world of difference between the task of counter-insurgency operations on urban streets and that of mounting or responding to long-range missile attacks on hostile states.

The latter point serves as a reminder of another drawback of the RMA analysis. It purports to offer a means to total victory, providing the opportunity for a universal war-fighting doctrine that, however, offers very little for the low-intensity conflict that was the combat norm during the Cold War, has become even more so subsequently and is likely to continue to be so. Moreover, the Americans have been able to take advantage of developments in doctrine and weaponry in order to retain their military lead, but it is necessary to remember that aggregate capability is not the same as capability or success in particular scenarios. In particular, there is a danger that the conviction of the value of high technology that lies at the centre of the RMA will serve as a cover for a failure to develop effective territorial control and counterinsurgency doctrines and practices, to conceive of a strategy for successful long-term expeditionary operations and to work within the difficult context of alliance policy-making and strategic control.

The threat from rogue states

Before turning to this type of warfare it is important to note that there are scenarios between that of territorial control/counterinsurgency warfare and that of conflict between two high-tech forces. In particular, the acquisition of advanced weaponry by so-called 'rogue' states creates the prospect of confrontation, even conflict, in which there is no equality of armament, but both sides rely heavily on technology or the threat of its use. In the late 1990s and early 2000s worries about developing Chinese, North Korean, Iranian, Iraqi, Libyan, Syrian and other long-range missile capability, including the growth of Chinese and Iranian submarine forces and the surprise test-firing by North Korea in August 1998 of a three-stage Taepodong I rocket over Japan into the Pacific, altered the balance of concern, led to a marked and public renewal in American interest in a comprehensive national missile defence scheme (NMD), and played a major role in the 2003 Gulf War.

Moves towards an American NMD led to discussion in Europe, Japan, Taiwan and other countries about the need for a similar system, or about how best to respond to American policies and their likely consequences for alliances with the USA and for relations with Russia and China. Thus, new missile-defence schemes may produce a new geopolitics. In Europe there has been concern about a 'decoupling' by the USA, as only it, not Europe, would be protected, in so far as protection is possible. It would require a more ambitious NMD to protect Europe, and it would also be much harder to persuade Russia to accept so extensive a revision of the Anti-Ballistic Missile (ABM) Treaty of 1972 (SALT 1). A similar threat to NATO has been seen in Canada, although it is likely that proximity will ensure that American defences cover most of its highly populated areas. Furthermore, the US Northern Command, established in 2003 for the defence of the continental USA and Alaska, includes Canada and Mexico within its scope.

The possession of advanced weaponry by rogue states enables them to offset, at least in part, the strength of the West's non-nuclear military capability, and provides both a valuable tool with which to intimidate local rivals and an important means for counterdeterrence. The latter includes removing the overhang of Western nuclear and non-nuclear power: a development which very much challenges American interests and exacerbates American concern that the failings of international bodies to maintain an orderly world will be mirrored by deficiencies in the American force profile. Thus the ability of the USA to protect its interests and allies seems limited unless its defensive system is enhanced, as well as redirected from a focus on the Soviet Union to include the threat from other states, the assumptions that underlie the much heightened expenditure on the military under George W. Bush and the attempts to reconfigure the armed forces.

Concern about the range of the missiles deployed in hostile states will exacerbate senses of vulnerability. These have a major impact on the politics of military preparedness. In essence, politicians and the public have a low tolerance of vulnerability and fear, which leads to demands for a secure and comprehensive defence system, but in reality none is likely to be both. Although satellite surveillance and real-time communications have improved, and will continue to do so, the speed of a missile attack and its likely combination with decoys and with satellite and communication disruption pose many problems. Decoys are a particular difficulty as outside the atmosphere there is no air resistance; hence objects of different weights and shapes behave in the same way and would be registered as similar by satellite surveillance. Furthermore, there is also the problem that warheads carrying chemical or biological substances would be designed to split, accentuating the problems of interception. More generally, there has been a major development in anti-satellite weaponry, tactics and doctrine, designed to counter satellite surveillance and interception. This is set to continue, challenging American hopes that a missile shield can offer invulnerability and encouraging consideration of plans that could not lead to weapons' deployment until a later date: for example for an airborne laser programme, which is designed to destroy enemy missiles in their boost phase, just after launching, rather than further into their trajectory. Possibly such a system would be less vulnerable to decoys, but its effectiveness is unclear.

As with Cold War atomic weaponry, we are dealing with the prospect of an attack for which there cannot be adequate defence trials, and, as was the case then, that does not prevent such trials, but also encourages planning for pre-emptive strikes. In the Cold War the danger of such attacks led to the ABM Treaty of 1972 which was designed to discourage a first strike by banning the construction of a defensive shield, and thus leaving the USA and the USSR vulnerable to a missile attack, so that there would be no effective defence against a counterstrike. The 1972 treaty served as the basis for further talks and encouraged restraint in the build-up of strategic-nuclear arsenals, but it is far from clear that a similar prospect of negotiation, agreement and restraint exists with those powers currently defined as rogue states. Furthermore, any process of negotiation would be far more complex as, due to nuclear proliferation and the development of missile systems, the number of participants would be greater than in the Cold War SALT talks.

It is appropriate to discuss the military threat from the rogue states with reference to the Cold War, because there are several elements in common, including an awareness of the need to plan for a war in which any success may itself be devastating to the victor as well as to the vanquished, even if the latter suffers far more. However, there are also important military differences: rogue states currently lack the intercontinental missile capability enjoyed by the Soviet Union and the USA during the Cold War, although they do increasingly possess a greater land mobility for their missiles. There were already important developments in this direction in the 1980s, and, now and in the future, the firing mechanism and infrastructure required for long-range missiles are lighter and more mobile than hitherto. This created problems in the first Gulf War – hunting the Iraqi Scuds – and these are likely to become greater in future conflicts.

As with Cold War atomic planning, but even more so given views on the maverick character of the leaders of rogue states, this situation will increase pressure for a first-strike option, and thus for reliable surveillance and analysis of enemy dispositions, capability and plans. The reliability of current surveillance will need to be enhanced, in particular to scan underground positions: a concern that has also encouraged the development of bunker-busting warheads. However, the processing and political evaluation of intelligence information are as, if not more, serious issues as the question of Iraqi weapons of mass destruction and their role in 2003 in the American and British decisions to attack indicated.

If confrontations and conflicts were to be a matter of missile and counter-missile preparations and exchanges alone, then the role in such operations of ground troops might appear limited: it would essentially be that of protecting missile positions, although there would also be a place for specialist deep-penetration units sent to attack such positions. It is unclear how far this is a realistic scenario. To turn to specifics, in the event of crises, rogue states such as North Korea or Iran would be put under Western (i.e. largely American) pressure by the deployment offshore of warships with offensive missile and anti-missile capability, and these ships would also serve as the base for airborne assault forces. If the latter can focus on attacking missile and communication centres, rather than on the difficult task of territorial control in the face of a hostile population, then they are likely to retain mobility and to make best use of their resources.

However, it is likely that opponents would locate mobile missiles in heavily populated areas, and also fire them from there.

Thus search-and-destroy missions will face powerful political and military constraints, while the military and political difficulties following the second Gulf War will encourage caution. These problems will not prevent preparations for further expeditionary warfare, but will encourage the drive for a technology of detection, location and interception in flight: a search that is not new, being one of the principal features of research in the latter period of the Cold War, when heat signatures from rocket blasts were detected by satellites, and the rocket's trajectory rapidly evaluated.

Earlier, during the Cold War, land-based intercontinental rockets ended American invulnerability, and this was followed by the development of hostile submarine forces capable of firing missiles into the USA from the American offshore. From that perspective, the rogue-states' scenario simply increases the number of possible missile-firers, and this will be particularly the case if these states acquire missile-launching submarines, for these overcome the constraints of national space, not least by vastly increasing an opponent's surveillance problems. Submarine forces are, however, more expensive to acquire and maintain than mobile land-rocket launchers, and the West has a well-developed capability in anti-submarine warfare. Furthermore, the small number of submarines that will be acquired ensures that detection and destruction are proportionately more of a problem for the power using submarines than the difficulties it faces in using land-based missiles. Thus, it is in the interests of the West that the entry costs to effective submarine capability remain high, and that submarine detection capability remains state-of-the-art.

Linked to this, it might seem best for the USA that the geopolitical concerns of rogue states focus on near neighbours, as that will ensure that these dominate their military tasking. While Syria, Iran and Libya might indeed desire an effective military capability able to threaten America, they will remain more concerned to acquire military assets that can threaten near-neighbours, such as each other and Israel. However, the range of US interests ensures that a geographical limitation of anxiety about the capability of these states is not possible. In part, this explains the threat from North Korea, because there the near neighbourhood includes a substantial US force in South Korea. In the Middle East, aside from American concerns about Israel and oil-production facilities, there are also, especially since the 2003 Gulf War, American forces that present a target. Similarly, British anxiety about Iraqi missiles in large part related to the threat to British military bases in Cyprus.

The diffusion of advanced weaponry will increase the opportunity cost of fighting rogue states. As was pointed out at the time, had Iraq or Serbia possessed atomic weaponry, then the response to the Gulf and Kosovo crises might well have been very different, and the same point could be made with reference to the successful international pressure on Indonesia in 1999 to withdraw its forces from East Timor. In addition, there are other forms of weaponry that would enhance the capability of rogue states. One such is satellite information, the absence of which kept the Iraqis in the dark about the flanking movement of the US Third Army during Operation Desert Storm in 1991.

A sense of vulnerability to universalist liberalism, in the form of renewed American assertiveness, has encouraged already strong pressures elsewhere to develop nuclear capability. In the late 1990s, both India and Pakistan publicly tested nuclear weapons and developed long-distance missiles capable of carrying such weapons. They were far from alone in such preparations and, in 2003, there was particular concern about Iran and North Korea, as, aside from their developing nuclear capability, each had long-range missiles. Thus in 2003 Iran conducted what it termed the final test of the Shahab 3 missile, first tested in 1998. With a range of 1,300 kilometres (812 miles), it is able to reach both Israel and US forces located in the region. The cascading nature of proliferation – the Iranian missile was based on a North Korean one – suggests that, in turn, the Iranians will spread the technology. As the pharmacology of terror greatly expanded, with developments in bacteriological and chemical weaponry, so the availability of long-range missiles became more of a military and diplomatic issue.

These changes suggest that the future prospect for political globalism is bleak. However much a state or a group of states might dominate the power stakes, as the USA now does, and is likely to continue doing, and however much diplomacy might resolve many problems, the costs of trying and failing to coerce a rogue state, such as Iran or North Korea, are likely to rise to a point that encourages caution. This would be not only a realpolitik scenario, but also the politics of prudence that most military leaderships are apt to encourage. On the other hand, as the experience of the Vietnam War suggests, politicians, while inexperienced about military matters, listen only to the advice that they wish to hear, and engage in promotion politics to ensure they receive this advice. Within the military, politicians can usually find those willing to offer it, and thus to advance or protect their own position.

Thus it is possible that the politics of prudence will be countered by the imperatives of commitment. A good example of this is likely to occur in the Middle East. A scenario can readily be discerned in which Egypt is taken over by a more anti-Western leadership, whether 'fundamentalist' or not, and another Arab–Israeli war breaks out, possibly as a result of instability in Palestine, Jordan, Lebanon or Syria, or of the ambitions of a post-Mubarrak government in Egypt. This might not go well for Israel, for although its military is far better equipped and trained than that of its Arab opponents the Israeli air force could be vulnerable to hand-held heat-seeking anti-aircraft weapons, and its armour to similar anti-tank weaponry. As another cause of conflict, the prospect of missile bombardment might draw the Israelis into a pre-emptive attack, as might the need to overcome their numerical inferiority by defeating their opponents in detail (i.e. separately), as they did in 1967. Whatever the cause, the political pressure on the Americans to intervene would be considerable. There would be prudential restraints, not least military commitments elsewhere (e.g. in Korea and maybe the Balkans), and the military advice might focus on the difficulty of the task, but it is likely that this would be ignored.

Israel might be regarded as an extreme case. Given its resonance in American domestic politics, and longstanding American commitment to it. Instead, it might be suggested that, had it occurred, a second Iraqi invasion of Kuwait at a time of major US commitments elsewhere (e.g. a serious confrontation with North Korea) might have been a more plausible litmus test of the USA's continuing willingness to act. Whatever the politics, however, the problem of stronger (than hitherto) armaments on the part of the non-Western combatant will recur, and from that perspective the Gulf War of 2003 was not very instructive as the Iraqi military, although powerful, had been weakened by over a decade of international sanctions.

Furthermore, this is not simply a matter of the diffusion of advanced weaponry to states outside the first rank, as there is also the problem of non-state organizations acquiring such weaponry. This is not new, and indeed was a feature of civil wars, but then these were weapons at the disposal of protostates, such as the Confederacy in the American Civil War, and of mass movements seeking to take over states, such as the Viet Minh. Now, advanced weapons can be acquired by movements that are far less numerous. The use of sarin nerve gas by Aum Shinrikyo, a Japanese sect, in an attack on the Tokyo Underground in 1995, showed that lethal bacteriological and chemical weaponry could be made and used by non-state organizations. The limited effectiveness of the sect's attacks, despite the considerable resources at its disposal, was less striking than the attempt to widen the terrorist repertoire. As with firearms and earlier weapons, state monopolization of the means of violence proved to be limited, and this has been underlined by discussion about how a crude or 'dirty' nuclear bomb could be fairly rapidly manufactured.

Such a situation did not mean that all weaponry could be easily replicated, or that it could be replicated in the quantities deployed by major powers; and al-Qaeda did not have nuclear, chemical, gas or bacteriological weapons when it attacked the USA in 2001. However, the potential capability of non-state players is a reminder of the danger of assessing and planning for conflict simply in state-to-state terms – a point underlined by the greater volatility of many of these players.

Counter-insurgency warfare

The spread of advanced weaponry can be related to the continuing problems of counterinsurgency warfare which will not diminish in future, and indeed may worsen, because the greater machinery-to-manpower ratio of the modern military is of scant value for the militarized policing functions that controlling territory in such circumstances entails. Thus in the 1990s the Angolan government used MiG–23 bombers, helicopter gunships, tanks and heavy artillery against the UNITA rebels, but found it difficult to contain their far-flung guerrilla attacks.

It is easy to be pessimistic about the likely consequences. In many respects, the brutality referred to as ethnic cleansing may become more common as states struggle to control hostile populations; while, as in Iraq in 2003, guerrillas seek to intimidate and terrorize alleged collaborators. Both processes were seen in the Angolan civil war, which finished only when the UNITA leader Jonas Savimbi was killed in 2002, although UNITA had already been seriously weakened when its supply route via Congo ended following the overthrow of its ally President Mobutu. Conversely, the problems that guerrilla operations encounter, including the supply of adequate munitions, may well encourage terrorism, instead of such operations; as well as lead to a level of non-cooperation that makes government difficult.

The nature of state response to sustained opposition will reflect ideologies and circumstances, but it is likely to be violent in those states that possess authoritarian political systems, a list that includes most of Africa and South-West Asia. Thus, the military will find the maintenance of government power its first obligation in many states. While in some countries this will be planned for with due care to civilian rights, and even with the use of non-lethal weaponry, in others there will be a rapid descent into brutal thuggery. The politicization of the military and the militarization of politics will be a synergy that it is all too difficult to overcome.

To turn to operational factors: the difficulties of dealing with a hostile population will be a factor both for 'developed' and for 'less-developed' military powers, and for both the difficulties in part reflect problems arising from the absence of any sharp differentiation of military from civilian opponents. Partly as a consequence, the value of successful manoeuvre warfare – the dislocation of opposing military structures, as in Iraq in 2003 – will be lessened by the need to address the problems of controlling territory – an issue that indicates the continued vitality of the defence at at least one level. This is important because not only Western technology but also its operational doctrine presupposes that success will flow from the seizure and maintenance of the initiative; Western forces are both trained and organized for this, and their structure and ethos are particularly appropriate for such operations. Indeed, Western military theory argues that the willingness to entrust decisions to low-level commanders encourages a fluidity that is the enabler of mobility, providing operational and tactical advantages over the forces of more centralized political–social systems. This may indeed work for the penetration battle, but war should involve a successful exit strategy as well as an effective entry one, and it is unclear that the military and political consequences of the long-term need to control territory have been adequately considered and integrated into planning – an issue that was readily apparent in Iraq in 2003.

Conclusions

Looking ahead, it is possible to see the character and processes of future conflict as very varied: this will represent a continuation of present trends, and it is difficult to foresee any revolutionary shift. The latter tend to be envisaged in terms of new weaponry, but, as already suggested, that confuses the means of conflict with its ends. It also neglects the extent to which innovations in weapons, tactics and strategy are rapidly diffused and/or quickly matched by counter-weapons, tactics and strategy. There is no reason to believe that the same will not continue to be the case, so that, for example, as satellites are developed both for surveillance and as weapons platforms, so anti-satellite capability will be enhanced across a range from killer satellites to ground-based weaponry, jamming and, more likely, evasion devices and practices.

The desire to control the battlefield, and reduce or manage risk, appeals to many facets of current Western culture, not least the absence of fatalism and the belief in planning. It also reflects the role of planning staffs, the nature of peacetime preparedness and the attempt to integrate planning into conflict as a continual process. However, it is less than probable that peacetime planning will 'work' as intended other than for offensive forces, with overwhelming power, launching very short conflicts. That does not mean that planning is without value, but rather that it cannot replace risk, and must not be used to encourage the notion of risk-free operations. Furthermore, it is necessary to understand the weakness of Western powers in managing exits from conflict. This is a function of excessive confidence in the 'total victory' that technological disparities are supposed to guarantee, Western incomprehension in negotiating with different worldviews, and cultural habits and institutional orientations within the military that can make it difficult to accept limited victory. A true military 'revolution' would entail, for example, a shift back towards widespread military service and/or a willingness to inflict and, still more, suffer heavy casualties. There are no signs of any such changes, and the enthusiasm with which commentators detect military revolutions needs to be contained as it leads to a failure to appreciate underlying continuities.

5 Conclusions

> We have regional dangers and the threat of aggression by
> hostile states against our friends and allies and interests in
> key regions. I'm speaking specifically, of course, about
> Southwest Asia and Northeast Asia. We have the possibility of
> regional instability and also failed states . . . We do not see
> or expect a regional power or a peer competitor for the next
> 10 or 15 years, but we have to prepare for that possibility
> certainly in the years beyond.
>
> (William Cohen, US Secretary of Defense,
> *Quadrennial Defense Review*, 1997)

We live in an age of dramatic displays of new capability,
and yet also of the persistence of traditional problems. For
example, in 1997, 500 American paratroopers embarked
in North Carolina, flew 6,700 miles and jumped into
Kazakhstan and Uzbekistan – an impressive display of
range; while modern weapons and platforms, and others
actively being developed, have capabilities that would
formerly have been the stuff of science fiction. At the same
time, in its fundamentals war changes far less frequently
and significantly than most people appreciate. This is not
simply because it involves a constant – the willingness of
organized groups to kill and, in particular, to risk death –
but also because the material culture of war (the weaponry
used and the associated supply systems) which tends
to be the focus of attention, is less important than its
social, cultural and political contexts and enablers. These

contexts explain the purposes of military action, the nature of the relationship between the military and the rest of society, and the internal structures and ethos of the military. Having and using high-tech weaponry, the focus of much discussion about the future of war, is not the same as winning particular wars, and in any case does not delimit the nature of conflict.

An awareness of the limitations of high-tech does not provide much guidance for the composition of future force structures, although it is valuable for an understanding of their likely use and capability, and thus for military doctrine. Instead, it is helpful to have an understanding of the destabilizing character of change. Change, indeed, is integral to the nature of the modern world, and it would be amazing if the twentieth century, which saw astonishing developments in capability and conflict, was succeeded by a century of stasis. There is no reason in modern culture or scientific capability to envisage such a shift, and the notion of modernity as a constantly changing presence and aspiration will continue. If anything, current trends indicate an acceleration of change as innovations in genetics, neuroscience and other fields proliferate through the rapid global communications system.

Acknowledging the role of change is not the same as stating the potency of all changes, but to focus on this role, and concentrating at this point on weaponry rather than doctrine, it is likely that at the high end of the conflict spectrum the weapons, platforms and weapons systems that dominate at the present moment will both remain central for probably two decades and then will be superseded. New weapons will need to combine firepower and mobility characteristics – the basic requirements of modern combat – but will try to lessen the frictions associated with use, as issues of durability, vulnerability and damage through use are all important.

To look further ahead, beyond a 40-year span (although these developments may occur earlier), is to consider the possibility of discovering and using new properties of matter, or at least of particular materials or ranges of spectrums. It is also likely that more effort will be devoted to creating effective means to disorientate the minds of opponents. Information-jamming will not be restricted to attacks on material systems, and the creation of precision disorientation weapons may be linked to future advances in the understanding of the chemistry and mechanics of the brain. The ability to affect these in a predictable fashion will be developed in medical science, and then there will be a rapid search for military application; indeed military research may well lead to developments in medical science. In turn, this will drive a search for counter-measures. Such suggestions imply that there will be further discussion about the 'laws of war': there may well be a sense that brain-interfering and sense-altering weaponry is unacceptable, but it is also likely to be seen as an alternative to physical destruction.

Looking further to the future, it is possible that cloning will be used to produce more 'disposable' soldiery, and/or that genetic manipulation will be employed to enhance capability. Genetic engineering, however, is likely to be more constrained than developments in the machanization of war that lead towards more effective robotic soldiers. In 2000 a Roboguard, a handgun mounted on a motorized arm, was produced in Thailand. The weapon can be programmed to fire automatically as soon as a target is selected: it does not require a human aimer, as it uses a video camera, infrared sensors and a laser sight. Such a weapon could easily be made mobile or given greater fire-power. It is probable that this invention will be pursued. The use of cloned or robotic soldiers will encourage research in technology designed to interfere with their control systems. Similarly, an awareness of the prospect of continued radical change in weaponry will affect investment in the platforms from which weapons are fired.

More generally, there will be a problem of continued investment demands. These will affect all powers, but their ability to respond will vary greatly, and this will help determine ongoing and subsequent absolute and relative capability. Issues will include not only wealth but also established patterns of taxation and expenditure, so that states that have heavy commitments in social welfare – for example, Germany – may find it difficult to match the expenditure of those that do not. Linked to this, but not coterminous, is the question of the relationship between the wealth of a society and the ability of the government to dominate and control resource use, which involves not only taxation levels but also the degree of state direction of the economy. Thus, those governments that may be able to spend the highest percentage of national resources on military preparations are authoritarian states, followed by non-authoritarian societies with a low level of social welfare and, lastly, those with a high level. In crude modern terms, this typology is represented by China, the USA and the EU.

In the past it was possible to argue that this was not too threatening a prospect, as authoritarian systems with their planning systems and arbitrary policies were inherently inefficient in terms of the requirements of capitalism for processes by which resources were best allocated in order to ensure productivity and growth. This involved more than economic processes, so that American society is far more flexible than autocratic societies, both in considering past and present, and in searching for advantage in the future. A society like the USA, which is open to talent and unhampered by a caste-like social structure, a rigid ideology or an autocratic government, should be better able to respond to the challenges of the future than a society affected by one or more of these characteristics. In military terms, this responsiveness includes being able to respond to the need for initiative in doctrine, strategy, tactics and weaponry. In both the Second World War and the Cold War the Americans were more successful than their opponents in encouraging and organizing a systemic productivity that provided a massive build-up of an effective military without jeopardizing the domestic economy; instead there was a positive synergy. It is unclear whether the future will be comparably benign, and this is crucial because US power politics depends on retaining a world-leading capability. As the Iraq 2003 crisis showed, American public culture with its moralism and self-confidence has little tolerance of the idea of restraints imposed by others, and although the search for national victory leaves many intellectuals uncomfortable it retains wide support in American public culture.

America is at present, and in the foreseeable future, the sole power able to set convincing limits to the aspirations of expansionist regimes and to give force to definitions of 'rogue' statedom. However much this capability is made less desirable to other powers by American unilateralism in goals and methods, it is important to the political health of the entire world. A move from American great power-dom to hegemony by another state – for example, China – or, more plausibly, to a condition in which there is no such hegemony, would mark more than simply a stage in the cyclical great-power theories of political scientists. The USA's lack of interest in territorial aggrandisement and the absence of racialism and religious intolerance in its public culture ensure that its global policy is relatively benign in terms of liberal anti-authoritarian standards. Such a comment may well surprise those used to the active or latent anti-Americanism of much world public debate on international relations – not least as seen in 2003. The argument that America's political and civil culture and economic strength and demands makes its global influence benign in a significant sense will for many be a counter-intuitive one; however, such arguments are important if we are to consider the complex and unpredictable context within which future conflicts will occur, and the difficulty of making judgements. Other counter-intuitive points that need stressing include the possibility that economic growth can increase the likelihood of conflict, and that 'losers' in war can reverse the verdict by other means.

Relative benignity does not mean that American power will not continue to support regimes and policies that fail to match up to the aspirations of American public culture, as it has done in Latin America and, in the case of its occupation policies, Israel. Yet this is in part a matter of the global representation of hegemonic power: it is not always, indeed usually, possible to choose allies that accord with one's ideals. Furthermore, as already indicated, American policy will test assumptions of international cooperation, and what that entails for world order. If the alternative to American power is seen as an authoritarian great power, or a widespread chaos exploited by rogue regimes, then it is relatively easy to see

America in benign terms, but this is far less the case if the alternative is seen in terms of a pluralistic international system cooperating through institutions, particularly the UN. From the latter perspective, the character of the USA as, in French terms, a hyperpower is a major problem, while the policies followed and aspirations voiced by the American government in 2001–03 are a challenge to the international order.

It is difficult, however, to envisage the replacement of American hegemony without either unsuccessful (for the USA) war, or the creation of another leading military power. If the former entailed an equivalent to the Vietnam War, it might be followed by a withdrawal from American interventionalism that left the world without a hegemonic power. This is more plausible, in the short or medium terms, than the creation of another power capable of matching the present position of the USA, but such a remark may well underrate future volatility in world politics, and a consequent willingness to support or accept a new hegemon.

The leading candidate is China, but it is unclear whether this would be a revisionist China seeking to remould the politics of part, or much, of the world, or, instead, a state willing to dominate while accepting the constraints of non-interventionism, specifically the sovereignty of other states. In addition, it is unclear whether the process by which China might achieve objectives at the expense of its near-neighbours may not involve a degree of conflict that ensures that Chinese great-power status is both distinctly militaristic and continually vigilant in the face of perceived revisionist tendencies on the part of states that have been intimidated or even defeated.

Chinese military power will continue to differ from that

of the USA for geopolitical, economic, political and social reasons. The first will ensure that naval power will be less important to Chinese force structure and doctrine: in the short term, the instability of Russia and Central Asia will accentuate this; but, in the long-term, lengthy borders with hostile neighbours will ensure it. This is a reminder that even symmetrical warfare does not involve conflict between identical forces: even if weapons are similar, force structures are not. There is a dependence of doctrine, technology and force structure on strategy, and of the latter on political concerns, that must not be forgotten. Social factors are also important. Chinese forces will continue to have a lower standard of living than their American counterparts, and thus fewer logistical demands, which will be important to their cost and usefulness, although the Chinese lack the lift enjoyed by the Americans. The demands of American forces and of their weaponry are such that the logistical tail is a major constraint, as well as providing a vulnerable link, open to malfunction, or to attack by advanced weaponry or by low-tech guerrilla opposition.

Whatever the nature of Chinese power in the future, it is likely that major states will continue to have to plan for symmetrical and asymmetrical conflict, and for high- and low-tech operations, but it is also necessary when looking to the future to accept that such categories are malleable and may indeed require continual redefinition. The last century and also the last decade underline the extent of unpredictability in human affairs. Repeatedly, predictions have been proved wrong, both about international relations and about domestic developments. There is no reason to believe that the future will be any different: on the contrary, the pace of change is likely to remain high and will probably accelerate, as the normative value of past and present arrangements declines in nearly all human societies.

Conceptual flexibility is important if a task- or threat-based approach to force structures and doctrine is taken, rather than – as is often tempting – a capabilities-based approach; in other words if the focus is on the tasks the military may be given and the threats they will confront, rather than simply building up their capability, in particular by acquiring advanced weapons systems and then basing force structures, plans and doctrine on these systems. The post-September 11 crisis posed this dilemma for the American military and state. Furthermore, the problem of preparing for the last war – a charge frequently made against the military – can in part be clarified and confronted by emphasizing the diversity of military tasks and the unpredictability of the manner in which these tasks present themselves in crises and conflicts.

The complexity of military tasking leads to an inevitable tension between politicians and public, who seek to have armed forces able to take on all tasks, and militaries, who point out the difficulty of achieving adequate flexibility with limited resources, necessitating the sophisticated management of priorities. Looking to the future, this prioritization will be most effective if it can escape the constraints of individual service interests, in short if overall force structures are more than the sum of compartmentalized services, but this flexibility will in large part depend on political support and direction. The tension between politicians and militaries played a role in the planning for the American-led attack on Iraq in 2003. Much of the debate on the rights and wrongs of this attack has focused on its legality or political prudence, but the military prudence also repays consideration. There was a serious

failure to consider the military consequences of a politically unwelcome outcome, in the shape of continued
resistance, and this left the American military in a
difficult position. As units are retained in Iraq for longer
than intended (and promised), so morale has fallen and
with it re-enlistments. The unpopularity of the occupation
also poses a question-mark against the continued validity
of the entire reservist system. The professional nature of
the American and British military greatly mitigates the
impact of poor political leadership, but the latter is a serious point, and from a variety of political and geopolitical
perspectives. Indeed, in light of the difficulty in achieving
acceptable political and military outcomes in Iraq, inter-
ventionists are going to find their case harder to make.
Doubtless, however, there will be promises of new technological advances that permit the overcoming of these and
other problems.

It is indeed possible to envisage many changes over the
next century. War might be dehumanized by entrusting
combat to computers, thus, apparently, taking mechanization to its logical conclusion. Alternatively, the vulnerability of human societies to environmental damage, and
of humans to disease, could be exploited in a systematic
form, expanding a possibility already seen with bacteriological warfare. In short, there is no reason to believe that
the capability of war for adaptation and major change will
diminish. Yet, alongside these ideas, it is more than likely
that standard aims will continue and that familiar problems will persist. How can states control dissident groups?
How can they guarantee security in an unstable world?
How can they use military capability to achieve their
objectives short of the unpredictable hazards of war? It is
difficult to feel that any of these issues will change. The

globalist aspirations of 1945 and 1990 seem defeated by the durability of differences within human society as much as by the continued centrality of the sovereign state and the lack across much of the world of stable civil societies. One prediction seems safe: talk of the obsolescence, even end, of war will prove misplaced, and will be mocked by the rictus on the face of the dead.

Index